Uncertainty Analysis of Experimental Data with R

Uncertainty Analysis of Experimental Data with R

Benjamin D. Shaw

CRC Press
Taylor & Francis Group

CRC Press is an imprint of the
Taylor & Francis Group, an **informa** business

A CHAPMAN & HALL BOOK

CRC Press
Taylor & Francis Group
6000 Broken Sound Parkway NW, Suite 300
Boca Raton, FL 33487-2742

First issued in paperback 2020

ISBN 13: 978-0-367-57339-3 (pbk)
ISBN 13: 978-1-4987-9732-0 (hbk)

Library of Congress Cataloging-in-Publication Data

Names: Shaw, Ben D.
Title: Uncertainty analysis of experimental data with R / Benjamin David Shaw.
Description: Boca Raton : CRC Press, 2017. | Includes bibliographical references.
Identifiers: LCCN 2016048268 | ISBN 9781498797320
Subjects: LCSH: Uncertainty (Information theory)--Textbooks. | Probabilities--Textbooks. | R (Computer program language)
Classification: LCC Q375 .S53 2017 | DDC 001.4/34028553--dc23
LC record available at https://lccn.loc.gov/2016048268

Visit the Taylor & Francis Web site at
http://www.taylorandfrancis.com

and the CRC Press Web site at
http://www.crcpress.com

I dedicate this book, with love, to my family.

Contents

1

Introduction

1.1 What Is This Book About?

This book covers methods for statistical and uncertainty analysis of experimental data of the types commonly encountered in engineering and also science. It is based on the author's experience teaching undergraduate and graduate engineering courses dealing with these subjects as well as topics the author found to be useful when confronted with analysis of data from experiments.

The primary focus is on analysis of results from experiments, which generally means that measurements of physical quantities have been done. An experiment could be as simple as measuring the length of a piece of string or as complicated as measuring the time-varying thrust produced by a rocket engine. The techniques for evaluating the statistics and uncertainties of these measurements are basically the same. If your measurement simply involves counting *discrete* events, then you can, in principle, accomplish this without error. For example, if you needed to count the number of people passing through a doorway, you could easily do this as long as you did not commit an error such as falling asleep. If you instead need to measure a quantity that can vary *continuously* such as length or temperature, then you generally cannot measure such a quantity exactly. For example, a length might be an irrational number such as the square root of 2, which has an infinite number of values after the decimal point. An exact measurement of such a quantity is clearly impossible. There will always be some uncertainty associated with a measurement, and it is a goal of this text to provide techniques that allow you to evaluate this uncertainty.

Uncertainties arise because of imperfections in measurements that lead to errors. A measurement error is the difference between the actual value of the quantity being measured and the measured value. Measurement errors are categorized as random (varying with time) or systematic (not varying with time) [1–3], and it is typically the case that errors from both categories are important in a measurement. We never know the exact value of a measurement error, so there is always doubt (uncertainty) about its value. Each error component in a measurement contributes to the overall measurement uncertainty via its own uncertainty component.

There are two popular approaches in use for evaluating component uncertainties. One approach [1,2] categorizes uncertainties that are a result of random errors as "random uncertainties" and uncertainties that are a result of systematic errors as "systematic uncertainties." The other approach [3] categorizes uncertainty components as Type A or Type B. Type A uncertainties are evaluated using statistical methods while Type B uncertainties are evaluated using other methods. Although these two methods differ in their implementation details, they should produce the same results for a given situation. The approach used by a particular researcher might depend on personal preferences or perhaps the uncertainty analysis methods required by an employer or a journal. The focus in this book will be on systematic and random uncertainties [1,2].

Uncertainty is sometimes characterized as the probable error of a measurement [2]. Another statement for uncertainty is that it is a nonnegative parameter characterizing the dispersion of values attributed to a *measurand* (the quantity being measured) [3]. Basically, knowing the uncertainty of a measurement will tell us something about an *interval* within which the true value of the measured variable is likely to be found. We typically will be interested in evaluating uncertainty intervals of the following form:

$$\text{uncertainty interval} = \text{middle value} \pm \text{uncertainty}.$$

For example, the uncertainty interval for a temperature measurement might be

$$T = 300 \pm 2\,\text{K}\left(95\%\right), \tag{1.1}$$

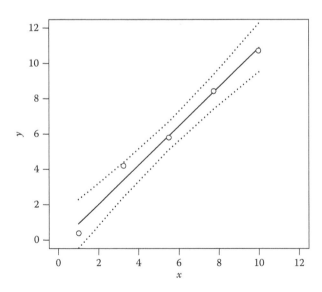

FIGURE 1.1
Experimental data (circles) with a curve fit (solid line) and 95% uncertainty band limits (dotted lines).

where the percentage represents a "confidence level." The confidence interval can be interpreted as meaning that, if we had performed this same experiment a large number of times, generating a new uncertainty interval for the data from each experiment, then about 95% of the uncertainty intervals we generated would contain the true value of the temperature.

A researcher will often want to investigate relationships between data. This can be accomplished by plotting data and also fitting equations. An example of this is shown in Figure 1.1. The circles correspond to experimental data, the solid line is a curve fit, and the dashed lines denote a 95% uncertainty band for this curve fit. Much of this text is focused on developing methods for curve fits as well as their uncertainties.

1.2 Units

Whenever we measure a physical quantity such as a length or a time, we need to relate that measurement to established standards. This means that we must use a specific system of units. We can regard units as the quantities to which physical quantities are referenced; e.g., length can be measured in feet, inches, miles, meters, centimeters, kilometers, light years, etc. A *base unit* is generally considered to be dimensionally independent of other base units. *Derived units*, such as force, energy, or pressure, are defined in terms of the base units. For example, one newton, which is a derived unit, is defined to be the force required to accelerate one kilogram with an acceleration of one meter per second squared ($1 \text{ N} = 1 \text{ kg m}/\text{s}^2$).

In this text we will use the metric system. Table 1.1 shows the base units for the current system [6]. A short list of derived units is presented in Table 1.2. The metric system is also named the Système international d'unités with the abbreviation SI and there are agreed-upon standards for the names and symbols of different quantities [7].

TABLE 1.1

Base Units of the Current SI System

Quantity	SI Name	SI Symbol
Mass	kilogram	kg
Length	meter	m
Time	second	s
Electric current	ampere	A
Thermodynamic temperature	kelvin	K
Amount of substance	mole	mol
Luminous intensity	candela	cd

TABLE 1.2

Some Derived Units for the SI System

Quantity	SI Name	SI Symbol	SI Units
Force	Newton	N	kgm/s^2
Velocity	—	—	m/s
Momentum	—	—	kgm/s
Energy	Joule	J	$kg\,m^2/s^2$
Pressure	—	—	$kg/(ms^2)$
Power	Watt	W	$kg\,m^2/s^3$

Except for the kilogram, the SI base units are presently defined in terms of experiments and procedures that, in principle, can be reproduced worldwide [8]. The kilogram is a platinum–iridium reference mass kept in a secure location in Paris, France. The second is defined as the time duration corresponding to 9,192,631,770 periods of the electromagnetic radiation emitted by cesium-133 atoms at 0 K. This radiation is specific to the transition between the two hyperfine levels of the cesium-133 ground state. A meter is the distance traveled by light in vacuum over the time of 1/299,792,458 s. The ampere is defined as the electrical current flowing through two parallel conducting wires of negligible diameter, spaced one meter apart and in vacuum, that produces a force of 2×10^{-7} N/m between the wires. The kelvin is 1/273.16 of the thermodynamic temperature of the critical point of water, where the water has a specific isotopic composition. The mole is the number of atoms in 0.012 kg of carbon-12. Finally, the candela is the luminous intensity in a given direction of a source of monochromatic electromagnetic radiation (light) of frequency 540×10^{12} Hz, where the radiant intensity in that direction is 1/683 watt per steradian. The radiant power of a typical household candle is about 1 cd.

The meter and the second are defined exactly. Since the standard for the kilogram is a piece of metal, however, the kilogram cannot be known exactly because we do not have instruments to make such an accurate measurement. Also, it has been established that the mass of the platinum–iridium reference mass has apparently *changed* by small amounts over time. The ampere, kelvin, mole, and candela cannot be determined exactly; i.e., there will always be some uncertainty associated with these measurements. The magnitude of this uncertainty can depend on factors such as operator skill, equipment characteristics, etc. Of course, it is possible to have a measurement apparatus calibrated, e.g., [9], but even with calibration, an instrument will have inherent uncertainties associated with it.

TABLE 1.3

Some Physical Constants and Their Relative Standard Uncertainties

Quantity	Value	Relative Standard Uncertainty
Universal gas constant	8.3144598 J/mol/K	5.7×10^{-7}
Avogadro's number	$6.022140857 \times 10^{23}$/mol	1.2×10^{-8}
Stefan–Boltzmann constant	5.670367×10^{-8} W/m^2/K^4	2.3×10^{-6}
Standard acceleration of gravity	9.80665 m/s^2	(exact)
Speed of light in vacuum	299,792,458 m/s	(exact)

1.3 Physical Constants and Their Uncertainties

Physical constants are present in almost all calculations in science and engineering, so it is crucial to have accurate numerical values for these constants. A source of values of physical constants is available [10]. Unless they are defined exactly, physical constants have uncertainties associated with them. For example, the standard acceleration of gravity and the speed of light in vacuum are defined exactly, but the universal gas constant is not. Current values (as of 2014) for a few physical constants and their relative standard uncertainties are listed in Table 1.3. The relative standard uncertainty of a quantity is the ratio of the standard deviation of the distribution of this quantity to its mean value. These uncertainties, which are typically very small, will be present in every calculation that uses these constants.

A new system of SI base units has been proposed and will likely be adopted within a few years. This new system will not change the operation of everyday measurement devices such as scales in a grocery store or a thermostat in your home, but it will, in principle, allow for replication of standards anywhere in the world. This includes the standard for the kilogram. The new base units are based on fundamental physical constants that are assumed to not change. Certain physical constants, e.g., Boltzmann's constant, will be assigned exact values and other quantities will be defined in terms of these constants.

1.4 Dimensionless Quantities

In science and engineering, it is often useful to express relationships between quantities in dimensionless form. This allows, e.g., experimental results obtained under one set of conditions to be used to predict what would happen under a different set of conditions. This works well if the conditions are *similar* in that the dimensionless values are the same. Dimensionless quantities always have the same values regardless of the unit system used.

Dimensional analysis techniques [11] can be used to define dimensionless quantities and to indicate a functional relationship between dimensionless variables. For example, consider a sphere moving in an incompressible Newtonian fluid. The drag force F acting on the sphere could be considered to be a function of the fluid–sphere relative speed V, the sphere diameter d, the fluid density ρ, and the fluid dynamic viscosity μ, leading to

$$F = f(\rho, V, d, \mu). \tag{1.2}$$

By using dimensional analysis, Equation 1.2 can be expressed as

$$C_D = g(\mathrm{Re}), \tag{1.3}$$

where the drag coefficient $C_D = F/(\rho V^2 \, \pi d^2/4)$ and the Reynolds number $\mathrm{Re} = \rho V d/\mu$ are dimensionless quantities. The functional relationship in Equation 1.3 would generally be determined from curve fits to experimental data, though various approximations can be used in the limits of Re being large or small relative to unity [11].

1.5 Software

We will use R [12] for data analysis. R is a software package that is useful for the analysis of experimental data. It is not the only software available for this purpose, but R is free and has become extremely popular over the past several years. There are many people worldwide who have contributed to the development of R. Many useful routines have been written in R and we will make use of some of these routines for analysis of data. R also has programming capabilities generally associated with a computer language, e.g., defining functions, making decisions, loops, etc. We will consider these topics as well. A goal of this text is to allow you to be able to write simple R scripts rather than always rely on code that someone else has written, but of course it can save a lot of time and effort if you are able to correctly use code that someone else has provided (assuming that this code works correctly).

We will cover aspects of R related to the topics covered in this book. If you would like more details than what is presented here, there are many references available (e.g., try a web search). However, some references that I personally found to be helpful are the (extensive) R text by Crawley [13], the R programming book by Matloff [14], and the statistics and data analysis books by Hothorn and Everitt [15], Radziwill [16], Vaughan [17], and Verzani [18].

The text by Hiebeler [19] has a comparison of R and the programming language MATLAB®. Braun and Murdoch [20] provide an introduction to statistical programming concepts using R. Stone [21] provides a tutorial introduction to Bayesian methods using R.

1.6 Topics Covered

We will first go over some aspects of R (Chapter 2), followed by a discussion of important statistics concepts (Chapter 3). We will then cover curve fitting (Chapter 4) and uncertainties (Chapters 5 through 7). The material in Chapters 5 through 7 is focused on an analytical Taylor series approach, which has its limitations. In Chapter 8 we will use computational (Monte Carlo) methods to provide more widely applicable approaches to analysis of uncertainty. The Bayesian approach is discussed in Chapter 9, and probability density functions are summarized in the Appendix.

Problems

P1.1 Discuss why it is thought that the mass of the platinum–iridium standard for the kilogram has been changing over time.

P1.2 What is the mass of one liter of liquid water at 4°C? Could you use this as a standard for one kilogram?

P1.3 Because the current SI system uses the speed of light in the definition of the SI base units for length and time, we can conceive of a system of units where all lengths are expressed in seconds. What would the SI units in the last column of Table 1.2 become in this case?

P1.4 The SI system has evolved as new scientific knowledge has been obtained. What were the original definitions of the base units for the meter, kilogram, and second for the SI system? What are the current definitions?

P1.5 Describe how the standard for the kilogram will be determined with the new system of SI base units.

P1.6 A saying sometimes used in cooking is "a pint's a pound the world around." How well does this saying apply to skim milk, water, and melted butter?

P1.7 Suppose you have been tasked with accurately measuring, with a thermocouple, the temperature of a liquid that is expected to be at 350 K. Using data from the website http://www.nist.gov/calibrations/, evaluate the minimum uncertainty you could expect to achieve. Assume that you can have your measurement system calibrated at the NIST.

P1.8   The current SI base units do not account for the effects of relativity. For example, if the temperature of an object increases, then the mass of this object also increases through Einstein's relation $E = mc^2$. (a) Suppose 1 kg of aluminum is initially at 300 K. What is the increase in mass if the temperature of the aluminum is raised to 400 K? (b) What is the increase in mass of 1 kg of steel if it is raised 1000 m in the earth's gravitational field? Assume that the steel's temperature does not change.

P1.9   The local acceleration of gravity can vary with time. What are the typical variations for a location for which you can find data? What are the implications of these variations?

P1.10 Describe how the standard for the ampere will be determined with the new system of SI base units.

References

1. ASME PTC 19.1-2013, Test uncertainty—Performance test codes, The American Society of Mechanical Engineers, New York, 2013.
2. H. W. Coleman and W. G. Steele, *Experimentation, Validation, and Uncertainty Analysis for Engineers*, 3rd edn., Wiley, Hoboken, NJ, 2009.
3. JCGM 101:2008, Evaluation of measurement data—Guide to the expression of uncertainty in measurement, GUM 1995 with minor corrections, International Bureau of Weight and Measures (BIPM), Sérres, France, 2008.
4. JCGM 101:2008, Evaluation of measurement data—Supplement 1 to the "Guide to the expression of uncertainty in measurement": Propagation of distributions using a Monte Carlo method, International Bureau of Weight and Measures (BIPM), Sérres, France, 2008.
5. B. N. Taylor and C. E. Kuyatt, Guidelines for evaluating and expressing the uncertainty of NIST measurement results, NIST Technical Note 1297, 1994 Edition (Supersedes 1993 Edition), National Institute of Standards and Technology, Gaithersburg, MD, 1994.
6. Bureau International des Poids et Mesures, Measurement units: The SI, 2016, http://www.bipm.org/en/measurement-units/ (accessed June 13, 2016).
7. R. A. Nelson, Guide for metric practice, *Physics Today* 56, August 2003.
8. NIST, Calibrations, 2016, http://www.nist.gov/calibrations/ (accessed June 24, 2016).
9. Bureau International des Poids et Mesures, Base units, 2016, http://www.bipm.org/en/measurement-units/base-units.html (accessed June 24, 2016).
10. NIST, The NIST reference on constants, units, and uncertainty, 2016, http://physics.nist.gov/cuu/Constants/index.html (accessed June 24, 2016).

11. H. G. Hornung, *Dimensional Analysis: Examples of the Use of Symmetry*, Dover Publications, Mineola, NY, 2006.
12. R Core Team. R: A language and environment for statistical computing. R Foundation for Statistical Computing, Vienna, Austria, 2014. http://www.R-project.org/ (accessed June 24, 2016).
13. M. J. Crawley, *The R Book*, 2nd edn., John Wiley & Sons, West Sussex, U.K., 2013.
14. N. Matloff, *The Art of R Programming: A Tour of Statistical Software Design*, No Starch Press, San Francisco, CA, 2011.
15. T. Hothorn and B. S. Everitt, *A Handbook of Statistical Analyses using R*, 3rd edn., CRC Press, Boca Raton, FL, 2014.
16. N. Radziwill, *Statistics (The Easier Way) with R: An Informal Text on Applied Statistics*, Lapis Lucera, San Francisco, CA, 2015.
17. S. Vaughan, *Scientific Inference: Learning From Data*, Cambridge University Press, Cambridge, U.K., 2013.
18. J. Verzani, *Using R for Introductory Statistics*, 2nd edn., CRC Press, New York, 2014.
19. D. E. Hiebeler, *R and MATLAB®*, CRC Press, New York, 2015.
20. J. W. Braun and D. J. Murdoch, *A First Course in Statistical Programming with R*, Cambridge University Press, Cambridge, U.K., 2007.
21. J. V. Stone, *Bayes' Rule with R: A Tutorial Introduction to Bayesian Analysis (Tutorial Introductions)*, Vol. 5, Sebtel Press, Sheffield, U.K., 2016.

2

Aspects of R

2.1 Getting R

R is compatible with Windows, Mac, and Linux operating systems. You can download the software from the website http://cran.r-project.org/. Although R can be used for analysis of many different kinds of data, we will cover aspects that are germane to analysis of scientific and engineering data. R documentation can be found at the website http://www.r-project.org/ (e.g., click on the "manuals" link). There are numerous books available as well, e.g., [1–7].

2.2 Using R

Rather than using a menu-based system, which can be limited in what it can do, we will use the command window that appears when R is started, as shown in Figure 2.1 for a Mac environment.

The fonts used in this book are as follows:

Arial font is used for text.

`Courier New font is used in the R command window.`

`Inputs are indicated as following the > symbol.`

`Outputs are indicated without the > symbol.`

`Errors will be introduced by "Error" or "Warning".`

You can customize your R environment to use different fonts.

The symbol > generally appears as the prompt in the command window. Note that # is the comment symbol, and any text following # on a line of input is ignored by R. The output from R is also shown below each command. The output can include numbers enclosed by square braces, e.g., [1],

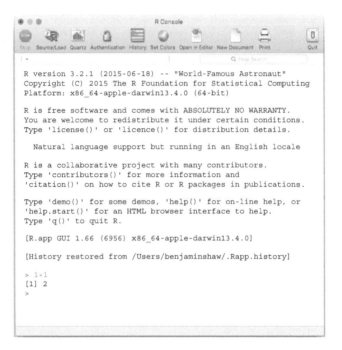

FIGURE 2.1
The R console window for a Mac environment.

at the beginning of each line of output—the number in braces simply gives the location of the first element in each line when the output is a list of elements. The cursor is typically a blinking vertical line.

In the R examples provided here, the command prompt (>) is included so that you can easily determine what you should type—do *not* type the command prompt when you enter commands. Also, the R examples we will use are often not the only possibilities for a certain type of data analysis. I have selected ways of doing things that I thought were the most transparent (at least to me).

2.3 Getting Help

R provides options for getting help with the commands help() and example(). You just need to fill in whatever you are asking about. For example, suppose you want to learn something about the plot() command. Just type the following commands and information should appear on your computer screen:

```
> help(plot)
> example(plot)
```

You can also get help by typing a question mark followed by whatever you want to be helped with:

```
> ?plot
```

2.4 Libraries and Packages

Although the R base software comes with many useful commands and routines, there will be occasions when you need to download other software. These "packages" can be downloaded from the Comprehensive R Archive Network (CRAN) using menu commands within R. The details are specific to whatever operating system you are using, so I will let you figure this out on your own.

Once you have downloaded a package, you load it into R using the library() command:

```
> library(shiny)        # load the package named shiny
```

You can have R list all of the packages that are available on your computer by using library() without any arguments. The packages that have actually been loaded are displayed with the search() command:

```
> library()        # generates a (long) list of packages
> search()         # displays loaded packages
  [1] ".GlobalEnv"          "package:locpol"    "tools:RGUI"
  [4] "package:stats"       "package:graphics"  "package:grDevices"
  [7] "package:utils"       "package:datasets"  "package:methods"
 [10] "Autoloads"           "package:base""
```

If you get an error message when you try to use a function,

```
> xFcn(x)
Error: could not find function "xFcn"
```

you probably need to load a necessary package into R.

2.5 Variables

Variables in R are case sensitive; i.e., x and X are different variables. A variable can be assigned a value using the assignment operator <-, which is a less-than symbol followed by a hyphen (minus sign).

For example, the variable named X can be assigned the value 5.2 by typing the following command:

```
> X <- 5.2
```

It is also possible to assign values using the equals sign (=), but this is not recommended because there are (rare) instances where an equals sign may not work correctly. Interestingly, the assignment operator can be used in the opposite direction, e.g.,

```
> 5.2 -> A
> A
[1] 5.2
```

2.6 Vectors

Variables in R are *vectorized* in the sense that a single variable can hold many values. For example, the variable Y can be assigned the vector (1, 2, 3, 4, 5) with the following command:

```
> Y <- c(1,2,3,4,5)
```

Note that c() is actually a function that combines (concatenates) its arguments, in this case the numbers 1, 2, 3, 4, and 5, into a vector.

  In some cases you may want to generate vectors with large numbers of elements, which is impractical to do manually. One way to do this is with the seq() command, which generates a vector of numbers with specified increments:

```
> seq(from = -0.2, to = 0.3, by = 0.1)
[1] -0.2 -0.1  0.0  0.1  0.2  0.3
```

Instead of using the by parameter to specify the increment, you can use length.out to specify the number of elements in the sequence. The operator : also generates a sequence of numbers. The command

```
> Z <- 1:50
```

generates the number sequence 1, 2, 3, ..., 50 and then stores the resulting vector in the variable Z. By typing a variable's name (and pressing return) we can see the values associated with the variable:

```
> Z
[1]  1  2  3  4  5  6  7  8  9 10 11 12 13 14 15 16 17 18 19
```

```
[20] 20 21 22 23 24 25 26 27 28 29 30 31 32 33 34 35 36 37 38
[39] 39 40 41 42 43 44 45 46 47 48 49 50
```

If we want to access a particular element of a vector, we do so using square brackets:

```
> Z[48]
48
```

If we want to remove a particular element of a vector, we can use square brackets but with a minus sign:

```
> Z[-48]
> Z
[1]    1  2  3  4  5  6  7  8  9 10 11 12 13 14 15 16 17 18 19
[20]  20 21 22 23 24 25 26 27 28 29 30 31 32 33 34 35 36 37 38
[39]  39 40 41 42 43 44 45 46 47 49 50
```

2.7 Arithmetic

R has the usual arithmetic operations found in any programming language (Table 2.1).

These operations can be used for calculations in an interactive manner as well as in scripts (which are described in Section 2.16), e.g.,

```
> 2^3                          # raise to a power
[1] 8
```

TABLE 2.1

Arithmetic Operations in R

Addition	+
Subtraction	–
Multiplication	*
Division	/
Exponentiation	^ or **
Modulus (remainder)	%%
Integer division	%/%

R does arithmetic *element by element* with vectors. Here is an example with multiplication of vectors:

```
> x <- c(1,2,3,4,5)
> y <- c(2,4,6,8,10)
> x*y                         # multiply element by element
[1]   2   8 18 32 50
```

Note that R *recycles* vector entries if an arithmetic operation is supposed to occur between two vectors of different lengths. What this means is that R will append copies of the shortest vector onto itself until it is the same length as the longer vector. The arithmetic operation is then performed. For this to work in a predictable manner, the longer vector should have a length that is an integer multiple of the length of the shorter vector. If it does not, then an error message may be generated. Here is some example code:

```
> x_short <- c(1,2,3,4)             # a 4-element vector
> x_medium <- c(7,8,9,2,3)          # a 5-element vector
> x_long <- c(2,4,6,8,10,12,14,16)  # an 8-element vector
> x_short+x_long
[1]   3   6   9 12 11 14 17 20
> x_short+x_medium
[1]   8 10 12   6   4
Warning message:
In x_short + x_medium :
  longer object length is not a multiple of shorter object length
```

If a vector is added to a number, each vector element is added to the same number:

```
> x <- c(1,2,3,4,5)
> x + 0.1
[1] 1.1 2.1 3.1 4.1 5.1
```

R treats a number as a vector of length 1 and then recycles it. This also applies to other arithmetic operations.

2.8 Data Frames

Data frames can be considered to be matrices but where the columns have names (headers). This allows for relatively easy access to specific columns of data simply by referencing the headers. We can create a data frame as follows, but in many cases, data frames are encountered when data files are imported into R.

Suppose we create the following three vectors:

```
> x <- 1:10
> y <- sqrt(x)
> z <- sqrt(y)
```

A data frame containing these vectors as columns can be created using the data.frame() command. Here, this data frame is associated with the object named W:

```
> x <- 1:10
> y <- sqrt(x)
> z <- sqrt(y)
> W <- data.frame(x,y,z)
> W
    x      y        z
1   1 1.000000 1.000000
2   2 1.414214 1.189207
3   3 1.732051 1.316074
4   4 2.000000 1.414214
5   5 2.236068 1.495349
6   6 2.449490 1.565085
7   7 2.645751 1.626577
8   8 2.828427 1.681793
9   9 3.000000 1.732051
10 10 3.162278 1.778279
```

Note that the column names are x, y, and z. Specific columns from a data frame can be accessed using the $ (dollar sign). This is accomplished by typing the name of the data frame, then $, then the column header. For example, we can access the z column using W$z:

```
> W$z
 [1] 1.000000 1.189207 1.316074 1.414214 1.495349 1.565085
 [7] 1.626577 1.681793 1.732051 1.778279
```

Here is the eighth data point in the z column:

```
> W$z[8]
 [1] 1.681793
```

2.9 Exporting Data

We can export data into a file on a computer. For example, suppose we create two new vectors, t_out and x_out:

```
> t_out <- 0:20
> x_out <- cos(t_out/3.14159)
```

We can export these data into a comma-separated values (csv) file with the following commands:

```
> # create a data frame named for.export
> for.export <- data.frame(t_out,x_out)
> # write the data into a csv file named "data"
> write.csv(for.export,file="data.csv")
```

The resulting file should now be available in the working directory.

2.10 Importing Data

R can access data files in several different formats; a common one is the csv format. The file we will use for illustration is named data.csv, which we created with the earlier code. This data file needs to be in the R working directory. You can set the working directory using menu commands in R or with the following command:

```
> setwd("~/filepath")  # fill in the filepath you need
```

Once the working directory has been set, you can read in the data with the read.csv() command:

```
> mydata <- read.csv("data.csv",header = TRUE,sep = ",")
```

The read.csv() command loads the data from the data file into the object we named "mydata." The entry header = TRUE tells R that the first entry in each column of data is actually a header, while sep = "," tells R that the entries in each row are separated by commas. The structure of mydata can be obtained with the str() command:

```
> str(mydata)
'data.frame':    21 obs. of 3 variables:
$ X     : int  1 2 3 4 5 6 7 8 9 10 ...
$ t_out: int  0 1 2 3 4 5 6 7 8 9 ...
$ x_out: num  1 0.95 0.804 0.578 0.293 ...
```

2.11 Internal Mathematical Functions

Many common mathematics functions are included in R (e.g., exponentials, sines, and cosines), e.g.,

```
> log(10)       # the log function is base "e" (i.e., it means ln)
[1] 2.302585
```

```
> log10(10) # you can also calculate base-10 logarithms
[1] 1
```

If you use a vector as the argument of a function, the function operates on each vector component separately:

```
> x <- c(1,2,3,4,5)
> z <- exp(-x)
> z
[1] 0.367879441 0.135335283 0.049787068 0.018315639 0.006737947
```

2.12 Writing Your Own Functions

It is relatively straightforward to write functions. The basic structure for a function we have named fcn is

```
> fcn <- function(x,y) {sin(x*y)}
```

This function takes two vectors, x and y, which have to be in the argument list of "function," and calculates another vector C defined as sin(x*y). The part that does the actual calculation is contained between the curly braces ({ }) and you can change this part to satisfy your own requirements.

Here is some code that uses this function:

```
> x <- 1:5
> y <- rnorm(length(x))
> z <- fcn(x,y)
> z
[1]   0.31229702 -0.50163372 -0.75735144   0.99787367   0.45329664
```

Note that x and y need to appear in the function argument list. The variables used in a function disappear after the function has finished its calculations, so we have assigned the result to another vector z to be able to use results from the function in other calculations.

2.13 Plotting Mathematical Functions

R can plot mathematical functions using the curve() command. For example, to plot the function sin(x) from x = 1 to x = 10, we could use the following command:

```
> curve(sin(x),from=1,to=10,col="red",xlab="x",ylab="sin(x)",n=101)
```

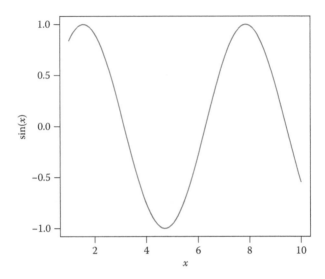

FIGURE 2.2
R plotting example.

The `col="red"` entry means that the line will be red, and the `xlab` and `ylab` entries specify the titles of the axes. The `"n=101"` entry specifies the number of points where the function is evaluated. The resulting output is shown in Figure 2.2. Note that the colors specified in the R code are rendering as black and white in this book because of the limitations imposed by black and white printing.

Another way to plot a function is to create a vector based on this function and then plot the vector, as shown in Figure 2.3:

```
> x <- seq(from=1,to=10,by=0.01)
> y <- sin(x)
> plot(x,y,xlab="x",ylab="sin(x)",col="blue",type="l")
```

2.14 Loops

Sometimes, an operation needs to be performed many times. A way to accomplish this is with a "`for`" loop. The structure of such a loop is illustrated here with an example where we add a sequence of numbers:

```
> ans <- 0
> for (i in 1:100) {
>       ans <- ans + i
> }
> ans
[1] 5050
```

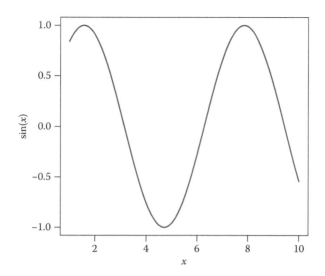

FIGURE 2.3
Another R plotting example.

In this example, the variable i is incremented by 1 as it goes from 1 to 100. Any commands between the curly braces ({ }) are executed every time i is incremented. The loop ends when i reaches a value of 100.

A "while" command can be used to execute a loop until a condition is met or a break command is executed. The earlier code using a for loop can be rewritten as a while loop in the following manner:

```
> ans <- 0
> i <- 0
> while (i <= 100) {
+       ans = ans+i
+       i = i+1
+ }
> ans
[1] 5050
```

The variable i is incremented by 1 and added to ans every time the loop is executed. The loop is exited when the condition i <= 100 is false.

2.15 Making Decisions

In certain cases you may need to apply a test to data and then make a decision based on the results of that test. This can often be done with an if–else statement, the structure of which is as follows:

```
if (expression) {statement 1} else {statement 2}
```

TABLE 2.2

Logical Expressions in R

Equal	==
Greater than or equal	>=
Less than or equal	<=
Greater than	>
Less than	<
Not equal	!=
Not x	!x
x OR y	x|y
x AND y	x&y
Test if x is TRUE	isTRUE(x)

If the "expression" is TRUE, then statement 1 is executed, and if the "expression" is FALSE, then statement 2 is executed. Whether an expression is TRUE or FALSE is determined by logical operators (Table 2.2).

Here is an example where we test whether the variable N is greater than zero and then set the value of the variable y based on the results of the test:

```
> N <- 10
> if(N > 0){y <- 1} else {y <- -1}
> y
[1] 1
```

The which() function can be used, in conjunction with logical expressions, to identify the elements of a vector that meet certain criteria. In the following example, which() is used to identify the indices of a vector's elements that are negative:

```
> x <- c(0,1,3,-4,-5,10,-10)
> which(x<0)
[1] 4 5 7
> x[which(x<0)]
[1]   -4   -5 -10
```

Note that which() provides the indices but not the actual values, so the line of code x[which(x<0)] is included to print out the negative values.

Similarly, the subset() function can be used to pull specific rows out of a data frame and save them in another data frame. Here is some code that creates a data frame with columns A, B, and C:

```
> A <- seq(from=0,to=10,by=1)
> B <- A-5
> C <- A^2
> my.data <- data.frame(A,B,C)
```

```
> my.data
      A   B    C
1     0  -5    0
2     1  -4    1
3     2  -3    4
4     3  -2    9
5     4  -1   16
6     5   0   25
7     6   1   36
8     7   2   49
9     8   3   64
10    9   4   81
11   10   5  100
> sub.my.data <- subset(my.data,B<0.1)
> sub.my.data
   A   B   C
1  0  -5   0
2  1  -4   1
3  2  -3   4
4  3  -2   9
5  4  -1  16
6  5   0  25
```

The subset function is then used to pull out the rows that have B < 0.1.

2.16 Scripts

In the examples so far, we simply typed code into the command window. This is useful, but it can be inefficient if we need to type long code sequences or if we are simply debugging some code we wrote and we need to run it many times. R has the capability to run scripts you write using a text editor that comes with R. Figure 2.4 is an image of some code I typed into the R editor. It was saved as a file named "script_example.R" on my desktop.

To run this code I made sure my working directory was the desktop and I then used the following command:

```
> source("script_example.R")        # make sure the name is in quotes
```

The output is shown in Figure 2.5.

You can also just copy the code in the editor, paste it into the command window, and run it by pressing return.

FIGURE 2.4
An example R script.

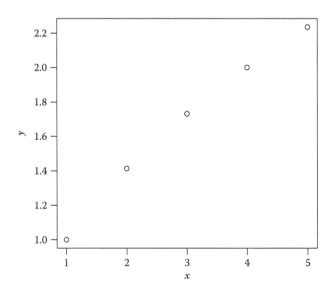

FIGURE 2.5
Output from an example R script.

2.17 Reading Data from Websites

R can access appropriately formatted data that are published on websites. For example, the NASA maintains a website of sunspot data. These data can be read into R with the following commands:

```
> sun_web <- "http://solarscience.msfc.nasa.gov/greenwch/spot_num.txt"
> sun_data <- read.table(sun_web,header=TRUE)
```

We can inspect the data using the head() and str() commands:

```
> head(sun_data)
  YEAR MON  SSN  DEV
1 1749   1 58.0 24.1
2 1749   2 62.6 25.1
3 1749   3 70.0 26.6
4 1749   4 55.7 23.6
5 1749   5 85.0 29.4
6 1749   6 83.5 29.2
> str(sun_data)
'data.frame':     3196 obs. of 4 variables:
 $ YEAR: int   1749 1749 1749 1749 1749 1749 1749 1749 1749 1749 ...
 $ MON : int   1 2 3 4 5 6 7 8 9 10 ...
 $ SSN : num   58 62.6 70 55.7 85 83.5 94.8 66.3 75.9 75.5 ...
 $ DEV : num   24.1 25.1 26.6 23.6 29.4 29.2 31.1 25.9 27.7 27.7 ...
```

So the data stored in sun_data are in the form of a data frame with column headers YEAR, MON, SSN, and DEV. If we want to, we can now generate plots using a command such as

```
> plot(sun_data$SSN,type="l")
```

For example, Figure 2.6 is a plot of the SSN (sunspot) data column.

2.18 Matrices and Linear Algebra

Matrices, i.e., rectangular arrays of numbers, can be constructed by defining vectors to be either rows or columns of the matrix using the cbind() or rbind() function. The cbind() function combines the vectors c(1,4,7) and c(2,5,9) as columns:

```
> m1 <- cbind(c(1,4,7), c(2,5,9))      # create matrix m1
> m1                                    # display matrix m1
```

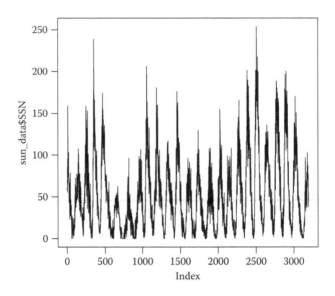

FIGURE 2.6
Sunspot data.

```
      [,1] [,2]
[1,]     1    2
[2,]     4    5
[3,]     7    9
```

In contrast, the rbind() function combines the vectors c(1,4,7) and c(2,5,9) as rows:

```
> m2 <- rbind(c(1,4,7), c(2,5,9))      # create matrix m2
> m2                                    # display matrix m2
      [,1] [,2] [,3]
[1,]     1    4    7
[2,]     2    5    9
```

Note that the rows and columns are denoted by the numbers in square brackets; e.g., [2,] denotes row 2 and [,3] denotes column 3. Specific row and column entries can be accessed using the [ ] operator:

```
> m2[2, 3]    # access a specific matrix element
[1] 9
> m1[2,]      # access all values in row 2 of matrix m1
[1] 4 5
> m2[,1]      # access all values in column 1 of matrix m2
[1] 1 2
```

Here are some basic matrix operations:

```
> m1 <- rbind(c(1,2),c(3,4))          # define matrix m1
> m2 <- rbind(c(5,6),c(7,8))          # define matrix m2

> m1                                   # display matrix m1
     [,1] [,2]
[1,]    1    2
[2,]    3    4

> m2                                   # display matrix m2
     [,1] [,2]
[1,]    5    6
[2,]    7    8

> m1%*%m2        # perform row-column matrix multiplication
     [,1] [,2]
[1,]   19   22
[2,]   43   50

> m1+m2                        # add matrices term by term
     [,1] [,2]
[1,]    6    8
[2,]   10   12

> m1-m2                        # subtract matrices term by term
     [,1] [,2]
[1,]   -4   -4
[2,]   -4   -4

> m1/m2                        # divide matrices term by term
           [,1]        [,2]
[1,] 0.2000000 0.3333333
[2,] 0.4285714 0.5000000

> m1*m2                        # multiply matrices term by term
     [,1] [,2]
[1,]    5   12
[2,]   21   32

> t(m1)                        # transpose
     [,1] [,2]
[1,]    1    3
[2,]    2    4

> dim(m1)                      # numbers of rows and columns
[1] 2 2
```

```
> det(m1)                # determinant
[1] -2

> diag(m1)               # diagonal entries
[1] 1 4

> eigen(m1)              # eigenvalues and eigenvectors
$values
[1]   5.3722813 -0.3722813
$vectors
              [,1]          [,2]
[1,] -0.4159736 -0.8245648
[2,] -0.9093767  0.5657675

> solve(m1)              # matrix inversion using LU decomposition
      [,1] [,2]
[1,] -2.0  1.0
[2,]  1.5 -0.5

> qr.solve(m1)           # matrix inversion using QR decomposition
      [,1] [,2]
[1,] -2.0  1.0
[2,]  1.5 -0.5

> b1 <- c(6,2)
> x1 <- solve(m1,b1)     # solve the equation m1%*%x1 = b1
> x1                     # display the solution x1
[1] -10    8
```

2.19 Some Useful Functions and Operations

Here are some lists of functions and operations we will use in this text. Not everything that is listed has been described so far, but I prepared this list in case you need something to refer to if you encounter something you have not been exposed to previously.

2.19.1 Data Frames

```
> attach()          # make data frame columns accessible by name
> data.frame()      # create a data frame
> dim()             # retrieve or set the dimension(s)
> fix()             # edit an R object such as a data frame
> head()            # display the first few rows of a data frame
> tail()            # display the last few rows of a data frame
> subset()          # select certain rows from a data frame
> W$x               # access column x in data frame W
```

2.19.2 Vectors

```
> X <- c(1,2,3,4,5)    # define a set of data as the vector X
> fivenum(X)           # print the data minimum, median, hinges, and maximum
> summary(X)           # print a summary of the data statistics
> X[2]                 # access the 2nd element of the vector X
> X[2:4]               # access elements 2 through 4 of the vector X
> X <- c(X,10.9)       # add a data point to the end of the vector X
> X[-6]                # remove data point number 6
> length()             # number of elements
> max()                # largest value
```

2.19.3 Probability and Statistics

```
> cor()                # correlation
> cov()                # covariance
> mean()               # average
> median()             # median
> min()                # smallest value
> prod()               # multiply all elements
> range()              # minimum and maximum values
> sample()             # draw a random sample
> sd()                 # standard deviation
> seq()                # generate a sequence of numbers
> sort()               # sort the data points from smallest to largest
> sum()                # add all the elements
> var()                # variance
> pchisq()             # chi-square cdf
> pnorm()              # normal cdf
> pt()                 # t cdf
> ptriangle()          # triangle cdf
> punif()              # uniform cdf
> qchisq()             # chi-square inverse cdf (quantile)
> qnorm()              # normal inverse cdf (quantile)
> qt()                 # Student's t inverse cdf (quantile)
> qtriangle()          # triangle inverse cdf (quantile)
> quantile()           # estimate data value below which P% of the data exist
> qunif()              # uniform inverse cdf (quantile)
> rchisq()             # random numbers from a chi-square distribution
> rnorm()              # random numbers from a normal distribution
> rt()                 # random numbers from a t distribution
> rtriangle()          # random numbers from a triangle distribution
> runif()              # random numbers from a uniform distribution
```

2.19.4 Plotting

```
> quartz()             # open a new plot window on a Mac
> windows()            # open a new plot window on a PC
> X11()                # open a new plot window on a Linux machine
> abline()             # draw a straight line
> boxplot()            # generate a box plot
> curve()              # plot a mathematical function
> density()            # generate a pdf curve
> identify()           # interactive plotting
> lines()              # add straight lines to join data points on a plot
> locator()            # interactive plotting
> pairs()              # generate a matrix of plots
> par()                # query or set graphical parameters
> plot()               # plot univariate or bivariate data sets
> points()             # add data points to a plot
```

2.19.5 Matrices and Linear Algebra

```
> cbind()              # combine vectors as columns
> det()                # determinant
> diag()               # diagonal entries
> eigen()              # eigenvalues
> qr.solve()           # matrix inversion using QR decomposition
> rbind()              # combine vectors as rows
> solve()              # matrix inversion using LU decomposition
> t()                  # transpose
```

2.19.6 Data, Functions, Libraries, and Packages

```
> builtins()           # list built-in objects
> data()               # list available data sets
> data(cars)           # load the data set named "cars"
> function() {}        # define a function
> library()            # display R packages installed on your computer
> library(shiny)       # load the R package named "shiny"
> search()             # display R packages that have been loaded
```

2.19.7 Various Other Functions and Operations

```
> example()            # show examples
> for () {}            # for loop
> help()               # display help information
> ls()                 # list workspace objects
> ls.str()             # list information about workspace objects
> rm(X,Z)              # remove objects X and Z from the workspace
> str()                # display the structure of an object
> source("sc.R")       # run the script "sc.R"
> which()              # find the indices for elements of interest
> while () {}          # while loop structure
```

Problems

P2.1 Create a variable X that contains the numbers 0.01, 0.02, …, 8.99, 9.00. Create the vector Y, which is the square root of X, as well as the vector Z, which is the logarithm base 10 of X. Plot Y vs. X as a red line and Z vs. X as symbols on the same graph.

P2.2 Create a data frame named my.data.frame that contains three columns of data (X, Y, and Z). The X column contains the numbers 1.00,

1.01, ..., 100.00, $Y = \sin(X)$, and $Z = \cos^2(Y)$. Plot Z and Y vs. X by accessing the columns in the data frame with $.

P2.3 Save the data frame from Problem 2 as a csv file named "my.data" on your desktop. Now import this csv file into R and check to make sure it has the data in the correct format.

P2.4 Using the subset() command, save all of the rows of data from the data frame created in Problem 2 where $Y > 0.5$. Plot Z and Y vs. X by accessing the columns in this data frame with $.

P2.5 Download sunspot data from http://solarscience.msfc.nasa.gov/greenwch/spot_num.txt, saving the data in a data frame. Use subsetting to pull out the data for the years 1950, 1952, and 1954 and save the subset data in another data frame. Examine the contents of this data frame to make sure they are what you expect.

P2.6 Create a function, e.g., using a while loop, that will calculate the sum

$$f(x,N) = x - \frac{x^3}{3!} + \frac{x^5}{5!} - \frac{x^7}{7!} + \cdots + \frac{x^N}{N!}$$

for arbitrary values of x (a real number) and N (a positive odd integer). This sum is actually $\sin(x)$ for N going to infinity. Plot $f(x,N)$ and $\sin(x)$ for $0 < x < 10$ for various values of N. Comment on how $f(x,N)$ converges to $\sin(x)$.

P2.7 Create a function that will calculate the sum

$$f(x,N) = \frac{1}{2\pi} \sum_{n=1}^{N} \frac{\cos(2\pi n x)}{n^2}$$

for arbitrary values of x and N. Plot $f(x,N)$ and $\sin(x)$ for $-2\pi < x < 2\pi$ for various values of N. Comment on whether $f(x,N)$ is converging to any particular form.

P2.8 Consider the following system of equations:

$$w - x + 2y - z = 2,$$
$$2w + x - 3y + 2z = 4,$$
$$w + x + 3y - 5z = 3,$$
$$w - 5x + 10y + z = 7.$$

Write this system in matrix form and solve for $w, x, y,$ and z using the linear algebra functions in R.

P2.9 Consider the following matrices:

$$m_1 = \begin{bmatrix} 1 & -1 & 2 & 4 \\ 3 & 4 & 2 & 5 \\ 2 & 7 & 1 & 6 \\ 2 & 2 & 5 & 8 \end{bmatrix}, \quad m_2 = \begin{bmatrix} 1 & 4 & 5 & 2 \\ 4 & 6 & 3 & 5 \\ 6 & 1 & 1 & 2 \\ 2 & 3 & 3 & 8 \end{bmatrix}.$$

Calculate the following quantities using the linear algebra functions in R: (a) the determinant of m_2, (b) the product of m_1 and m_2, and (c) the eigenvalues of m_1.

P2.10 The exponential of a square matrix S is defined as follows:

$$e^S = S^0 + S^1 + \frac{S^2}{2!} + \frac{S^3}{3!} + \cdots + \frac{S^n}{n!} + \cdots.$$

Write a function in R that will take an arbitrary square matrix and evaluate this series out to N terms. Note that S^0 is the identity matrix with the same dimensions as S. Determine the result if $N = 10$ and

$$S = \begin{bmatrix} 1 & 4 & 5 \\ 2 & 4 & 6 \\ 3 & 5 & 6 \end{bmatrix}.$$

References

1. M. J. Crawley, *The R Book*, 2nd edn., John Wiley & Sons, West Sussex, U.K., 2013.
2. N. Matloff, *The Art of R Programming: A Tour of Statistical Software Design*, No Starch Press, San Francisco, CA, 2011.
3. T. Hothorn and B. S. Everitt, *A Handbook of Statistical Analyses Using R*, 3rd edn., CRC Press, Boca Raton, FL, 2014.
4. N. Radziwill, *Statistics (The Easier Way) with R: An Informal Text on Applied Statistics*, Lapis Lucera, San Francisco, CA, 2015.
5. S. Vaughan, *Scientific Inference: Learning from Data*, Cambridge University Press, Cambridge, U.K., 2013.
6. J. Verzani, *Using R for Introductory Statistics*, 2nd edn., CRC Press, New York, 2014.
7. D. E. Hiebeler, *R and MATLAB®*, CRC Press, New York, 2015.

3

Statistics

In this chapter we consider aspects of statistics that are relevant to data analysis. If you need more details than are presented here, there are many statistics textbooks available. Two textbooks that focus on using R are by Verzani [1] and Radziwill [2]. The R text by Crawley [3] also covers many aspects of statistics. Hothorn and Everitt [4] cover advanced topics related to using R for statistical analysis.

3.1 Populations and Samples

A population is the totality of possible values that a particular physical value can attain in an experiment, whereas a sample is a subset of the population. In general, it is the population we are interested in. Variables such as the mean or the median, which summarize some aspect of the population or the sample, are termed *statistics*.

For quantities that vary continuously over some range, the population is infinitely large even if the population extremes are known. For example, how many possible values of temperature are there in the temperature range of 0–100 K? Of course, there are an infinite number of values in this range (or any finite temperature range). Because we cannot measure all of these values, we take a sample composed of a finite number of measurements and use this sample to infer characteristics about the population. There is also the possibility that the population is composed of discrete elements, though this is not common in physical science and engineering, and we will not consider this here.

When we are analyzing data (i.e., samples), we are usually interested, at least in the beginning, in the following:

1. Calculating the mean, median, standard deviation, and variance of a sample
2. Evaluating covariances and correlations between variables
3. Visualizing the data to get a sense of its distribution and trends
4. Using the sample to estimate statistics of the population
5. Identifying potential outliers

We consider these topics, and others, in the following.

3.2 Mean, Median, Standard Deviation, and Variance of a Sample

The sample mean (arithmetic average) $\bar{x}$ is defined as

$$\bar{x} = \frac{1}{N}\sum_{i=1}^{N} x_i, \tag{3.1}$$

where
   N is the number of data points
   x_i is data point i

The median x_m is found by ordering the data from largest to smallest and then defining x_m to be the middle data point if N is an odd number or the arithmetic average of the two middle data points if N is even, i.e.,

$$x_m = x_{(N+1)/2}, \quad N \text{ odd,} \tag{3.2}$$

$$x_m = \frac{x_{N/2} + x_{N/2+1}}{2}, \quad N \text{ even.} \tag{3.3}$$

The variance of a sample, s_x^2, is defined as

$$s_x^2 = \frac{1}{N-1}\sum_{i=1}^{N}(x_i - \bar{x})^2, \tag{3.4}$$

and the sample standard deviation s_x is simply the square root of the variance:

$$s_x = \left(s_x^2\right)^{1/2} = \left(\frac{1}{N-1}\sum_{i=1}^{N}(x_i - \bar{x})^2\right)^{1/2}. \tag{3.5}$$

The standard deviation is a measure of the average spread of the data about the mean. It is to be noted that in some texts, the $N-1$ terms in these equations are replaced with N. The approach to follow is to use $N-1$ if you are going to use s_x and s_x^2 to characterize the sample, which is what we do here.

For illustration, we will calculate these basic statistics using R's built-in functions for the following data set:

```
> x <- c(2.3,4.2,3.2,4.1,1.1,5.4,3.3,4.4,3.7,2.7)    #data
> mean(x)                    # sample mean
[1] 3.44
> median(x)                  # sample median
[1] 3.5
> var(x)                     # sample variance
[1] 1.471566
> sd(x)                      # sample standard deviation
[1] 1.213077
```

3.3 Covariance and Correlation

Covariance and correlation are important when we want to assess whether the variations in two variables are correlated, i.e., whether they move up and down together in some sense. If we have two data sets $\{x_1, x_2, ..., x_N\}$ and $\{y_1, y_2, ..., y_N\}$, the covariance s_{xy} is defined as

$$s_{xy} = \frac{1}{N-1} \sum_{i=1}^{N} (x_i - \bar{x})(y_i - \bar{y}) \tag{3.6}$$

and the correlation coefficient ρ_{xy} is defined as

$$\rho_{xy} = \frac{s_{xy}}{s_x s_y}. \tag{3.7}$$

Here, each data set has the same number of elements and we have not reordered them in any way as this can influence the level of correlation. For example, sorting both x and y from smallest to largest will force them to be correlated. The correlation coefficient is always in the range $-1 \leq \rho_{xy} \leq 1$. Two data sets are perfectly correlated if $\rho_{xy} = +1$ or -1 and completely uncorrelated if $\rho_{xy} = 0$, which means that $s_{xy} = 0$. If ρ_{xy} is "close" to +1 or −1, the data are mostly correlated, and if ρ_{xy} is "close" to 0, the data are mostly uncorrelated. The covariance of a data set with itself is the variance of this data set, i.e., $s_{xx} = s_x^2$. Because a data set is perfectly correlated with itself, it is always the case that $\rho_{xx} = \rho_{yy} = 1$.

In R, the covariance and correlation of two data sets are calculated using the functions cov() and cor(), respectively. Here is an example where we plot a data set y as a function of x (Figure 3.1):

```
> x <- seq(from=0,to=10,length.out=25)
> y <- x + rnorm(length(x))
```

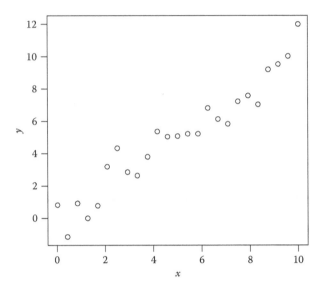

FIGURE 3.1
Plot of a data set.

```
> plot(x,y)
> cov(x,y)
[1]  9.478072
> cor(x,y)
[1]  0.9583428
```

The y data are noisy and we calculate the correlation coefficient as being $\rho_{xy} = 0.96$. This indicates that the data are well correlated, which is also evident in the plot. Always remember, though, that *correlation does not imply causation*. There could be another variable that is causing x and y to vary.

3.4 Visualizing Data

3.4.1 Histograms

A histogram is useful for visualizing the general distribution of a univariate data set, i.e., a data set in which there is only one variable, such as the length of an object. When a histogram is constructed, the data are sorted into rect-angularly shaped "bins," with the height of a bin being proportional to the frequency (i.e., the number of data points within that bin). In R, histograms are constructed using the command hist().

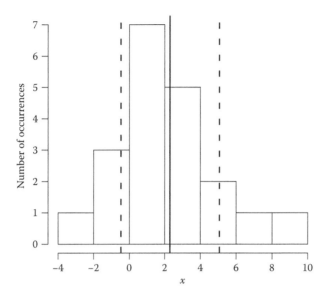

FIGURE 3.2
A histogram of a data set. The solid vertical line shows the data mean and the dashed lines show one standard deviation on either side of the mean.

For illustration, we will use the following data set:

```
> > x <- c(0.83,0.59,-0.05,1.44,4.59,1.08,9.12,-2.20,7.00,-0.15,1.90,-0.79,1.07,
  3.29,2.31,0.25,5.73,3.68,3.71,2.54)
```

Figure 3.2 shows a histogram for this data set obtained using the following code:

```
> hist(x,xlab="x",ylab="Number of Occurrences",main="")
> abline(v=mean(x),col="black",lw=2)
> abline(v=mean(x)-sd(x),col="black",lw=2,Hy="dashed")
> abline(v=mean(x)+sd(x),col="black",lw=2,Hy="dashed")
```

We added vertical lines with the abline() function to mark the data mean (solid line) and one standard deviation on either side of the mean (dashed lines). The bin widths are automatically set in this example, but there are options that allow you to set your own bin widths.

3.4.2 Box Plots

Box plots are also useful for visualizing univariate data distributions. An example shown in Figure 3.3 was created using the following code:

```
> x <- c(0.83,0.59,-0.05,1.44,4.59,1.08,9.12,-2.20,7.00,-0.15,1.90,-0.79,1.07,3.29,2.31,
  0.25,5.73,3.68,3.71,2.54)
> boxplot(x,range=1.5,horizontal=TRUE,xlab="x")
```

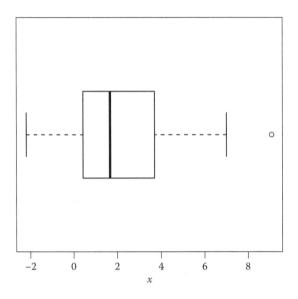

FIGURE 3.3
A box plot.

The "box" in a box plot generally contains the middle 50% of the data. The median is marked using the thick line inside the box and the sides of the box denote data quartiles. The whiskers extend out a user-specified distance, i.e., some number times the interquartile range (IQR), to either side of the box. Box plots can also be useful for identifying data points that can be considered potential outliers by marking points that are outside the whiskers.

The term "range=1.5" specifies how many IQRs to extend the whiskers from the sides of the box. In this example, a possible outlier is indicated by the circle. Using "range=0" extends the whiskers to the data extremes. The term "horizontal=TRUE" draws the box plot horizontally.

We can print numerical data corresponding to this box plot with the fivenum() command:

```
> fivenum(x)
[1] -2.200   0.420   1.670   3.695   9.120
```

The first and last entries in the output are the minimum and maximum values, respectively, and the middle three values are the quartiles (hinges) Q_1, Q_2, and Q_3, where Q_2 is the sample median.

A useful aspect of box plots is that they provide a way to visually compare data sets simply by putting more than box plot on the same plot. To show

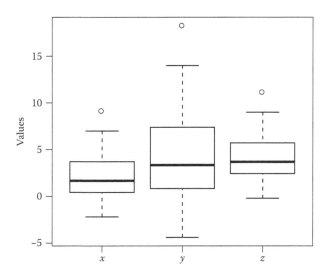

FIGURE 3.4
Comparison of three data sets with box plots.

this, we will create a data frame with three columns of data and then use the `boxplot()` command on the data frame:

```
> x <- c(0.83,0.59,-0.05,1.44,4.59,1.08,9.12,-2.20,7.00,-0.15,1.90,-0.79,1.07,3.29,
  2.31,0.25,5.73,3.68,3.71,2.54)
> y <- 2*x
> z <- x+2
> W <- data.frame(x,y,z)
> boxplot(W,range=1.5,horizontal=FALSE,ylab="Values")
```

The resulting box plots are shown in Figure 3.4. We can readily see how the data medians and spreads vary among these data sets. Note that the "horizontal=FALSE" term causes the box plots to be displayed vertically.

3.4.3 Plotting Data Sets

We can generate a plot of univariate data using the `plot()` function and then connect the data points (if we want to) with straight lines using the `lines()` function:

```
> x <- 1:20              # generate some data
> x <- exp(-x/3)         # generate some data
> plot(x)                # plot the data points
> lines(x)               # connect the data points
```

The resulting plot is shown in Figure 3.5.

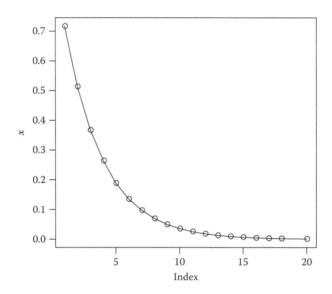

FIGURE 3.5
Plot of a univariate data set.

To plot a bivariate set of data, i.e., a data set in which y is a function of another variable x, we can use the `plot()` command to generate a plot (Figure 3.6):

```
> x <- c(1,2,3,4,5,6,7,8,9,10)              # vector of x values
> y <- c(2,5,3,7,8,11,2,3.4,2.9,1.09)       # vector of y values
> # generate a scatter plot
> plot(x,y,xlim=c(0,10),ylim=c(0,15),xlab="x",ylab="y")
> lines(x,y,col="black")      # connect the points with black lines
```

This `plot()` command plots the data points as circles (as the default) and the limits for the axes are set by the `xlim=c(0,10)` and `ylim=c(0,15)` entries. Black lines are then added to connect the data points using the `lines(x,y,col="black")` command.

If we want to plot more than one data set, e.g., we want to plot both y and z as a function of x, we can do this by first plotting y with `plot()` and then adding z with the `points()` and `lines()` commands (Figure 3.7):

```
> x <- c(1,2,3,4,5,6,7,8,9,10)              # vector of x values
> y <- c(2,5,3,7,8,11,2,3.4,2.9,1.09)       # vector of y values
> z <- c(1,3,5,7,3,4,6,8,9,2)  # vector of z values
> # generate a scatter plot of y vs. x
```

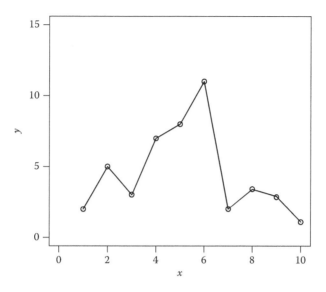

FIGURE 3.6
Plot of a bivariate data set.

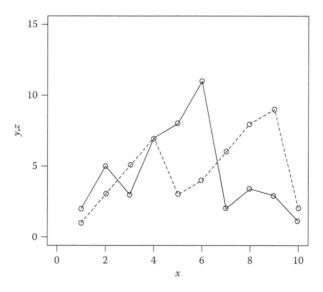

FIGURE 3.7
Plot of two bivariate data sets.

```
> plot(x,y,xlim=c(0,10),ylim=c(0,15),xlab="x",ylab="y,z")
> lines(x,y,col="black")           # connect the points with black lines
> points(x,z,col="black")          # plot z vs. x points
> lines(x,z,col="black",Hy="dashed")   # connect the x-z data points
```

3.4.4 Some Plotting Parameters and Commands

You can control the plot symbol and the line type via some simple commands. Here is the code to show 20 of the available plot symbols:

```
> plot.new()
> par(usr=c(-1,21,0,1))
> for (i in 0:20) {
       points(i,0.5,pch=i,cex=2,col="black")
       text(i,0.45,i)
       }
```

The resulting output is shown in Figure 3.8.

The commands to control the plot symbols are in points(), where pch sets the shape, cex sets the size, and col sets the color.

The type of line can be controlled in a plot using the lty parameter that can be used in either the plot() or the lines() command. In Figure 3.9

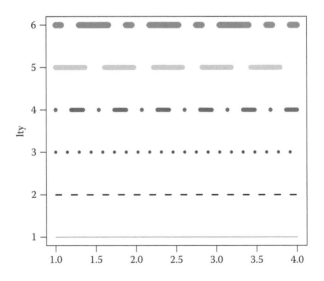

FIGURE 3.8
Some plot symbols in R.

FIGURE 3.9
Example of various lines that can be plotted.

we have plotted six lines of different types, colors, and widths created with the following code:

```
> plot(c(1,4),c(1,1),type="n",ylim=c(1,6),xlab="",ylab="lty")
> lines(c(1,4),c(1,1),lw=1,col="magenta",lty=1)
> lines(c(1,4),c(2,2),lw=2,col="black",lty=2)
> lines(c(1,4),c(3,3),lw=3,col="blue",lty=3)
> lines(c(1,4),c(4,4),lw=4,col="red",lty=4)
> lines(c(1,4),c(5,5),lw=5,col="green",lty=5)
> lines(c(1,4),c(6,6),lw=6,col="purple",lty=6)
```

The line type is indicated on the vertical axis. The line width is specified with the option `lw` and the line color with `col`. Because this book is printed in black and white, the colors in Figure 3.9 render as gray scale.

3.5 Estimating Population Statistics

We generally obtain samples so that we can learn something about the statistics of a population. Unfortunately, the best we can usually do is to provide estimates of population statistics. A *point estimate* is a value such as s_x that we hope is close to σ, while an *interval estimate* is a range within which we think a statistic such as σ might lie. Interval estimates are also called confidence intervals if there are no systematic errors or uncertainty intervals if there are systematic errors. In the following, we present a few methods for evaluating confidence intervals. We will not go into the detailed derivations of the methods but rather will focus on showing you how to use the methods with R code.

3.5.1 Confidence Interval for the Population Mean Using Student's *t* Variables

A $P\%$ confidence interval for the population mean can be estimated using

$$\mu = \bar{x} \pm t_{\nu,P} s_x / \sqrt{N}\ (P\%), \tag{3.8}$$

where
  $t_{\nu,P}$ is Student's *t* variable
  N is the number of data points
  $\nu = N - 1$ is the number of "degrees of freedom"

There is a *built-in* function named `t.test()` that we can use for this calculation. We will apply this function to a set of data as follows:

```
> x <- c(2.3,4.2,3.2,4.1,1.1,5.4,3.3,4.4,3.7,2.7,3.0,1.9,7.9,4.6,3.4)
> P <- 0.95    # confidence level (we select it)
```

```
> t.test(x,conf.level=P)
        One Sample t-test
data:   x
t = 8.8654, df = 14, p value = 4.062e-07
alternative hypothesis: true mean is not equal to 0
95 percent confidence interval:
 2.789707 4.570293
sample estimates:
mean of x
      3.68
```

The 95% confidence interval for the population mean is thus

$$2.79 \le \mu \le 4.57 \ (95\%)$$

or, equivalently,

$$\mu = 3.68 \pm 0.90 \ (95\%).$$

3.5.2 Confidence Interval for the Population Variance Using Chi-Square Variables

A $P\%$ confidence interval for the population variance can be estimated using

$$\frac{vs_x^2}{\chi_{v,\alpha/2}^2} \le \sigma^2 \le \frac{vs_x^2}{\chi_{v,1-\alpha/2}^2} \ (P\%), \tag{3.9}$$

where
χ^2 is the chi-square variable
$\alpha = 1 - P$ is the "level of significance"
$v = N - 1$ is the number of degrees of freedom

There is a built-in function named sigma.test() in the package "TeachingDemos" that we can use for this calculation:

```
> library(TeachingDemos)
> x <- c(2.3,4.2,3.2,4.1,1.1,5.4,3.3,4.4,3.7,2.7,3.0,1.9,7.9,4.6,3.4)
> P <- 0.95           # the selected confidence level
> sigma.test(x,conf.level=P)
        One sample Chi-squared test for variance
data:   x
X-squared = 36.184, df = 14, p value = 0.001958
alternative hypothesis: true variance is not equal to 1
95 percent confidence interval:
 1.385354 6.428453

sample estimates:
var of x
2.584571
```

The confidence interval for the population variance is

$$1.39 \leq \sigma^2 \leq 6.43 \, (95\%)$$

and the standard deviation (found by taking square roots) has the confidence interval

$$1.18 \leq \sigma \leq 2.53 \, (95\%).$$

3.5.3 Confidence Interval Interpretation

An interpretation of a $P\%$ confidence interval is that if we performed a large number of experiments and calculated a confidence interval for some statistic, e.g., μ, for each experiment, then approximately $P\%$ of these confidence intervals would actually contain μ. Unfortunately, though, we would not know *which* confidence intervals actually contain μ.

Here is a simulation to illustrate this. We draw random samples from a population and calculate confidence intervals for the population mean. These confidence intervals are then plotted in Figure 3.10. The dashed line in the plot is the true mean of the population, the sample means are denoted by the circles, the gray confidence intervals include the true mean, and the black confidence intervals do not. Note that the fraction of confidence intervals that include the population mean is approximately the confidence

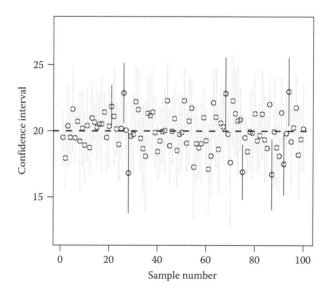

FIGURE 3.10
Confidence intervals from a set of repeated calculations.

level P. In this simulation, the fraction of confidence intervals that contained the mean was 0.94 and the confidence level was $P = 0.95$. Here is the code used to generate the plot:

```
> x_mean <- 20                        # population mean
> x_sd <- 4                           # population standard deviation
> x <- rnorm(1000000,mean=x_mean,sd=x_sd) # create a large population to
                                           sample from
> P <- 0.95              # confidence level
> n <- 10                # number of data points in a sample
> N <- 100               # number of samples
> ci_min <- numeric(N)   # vector to contain lower confidence interval limits
> ci_max <- numeric(N)   # vector to contain upper confidence interval limits
> I <- 1:N               # vector containing the sample numbers (indices)
> frac <- 0 # initialize the fraction of samples containing the true mean
> # generate the confidence intervals
> for (i in 1:N) {
+     x_sample <- sample(x,n)       # random sample i of n data points
+     a <- t.test(x_sample,conf.level=P)  # t test with a confidence level P
+     ci_min[i] <- min(range(a$conf.int)) # lower confidence interval limit for
                                            sample i
+     ci_max[i] <- max(range(a$conf.int)) # upper confidence interval limit for
                                            sample i
+     if (x_mean >= ci_min[i] && x_mean <= ci_max[i]) frac <- frac + 1/N
          # update frac
+ }
> # plot the middle value for each confidence interval
> quartz(width=5.5,height=5.5)
> plot(I,(ci_min+ci_max)/2,xlim=c(1,N),ylim=c(x_mean-2*x_sd,x_mean+2*x_
   sd),xlab="Sample Number",ylab="Confidence Interval",cex=1)
> # plot the confidence intervals
> for (i in 1:N) {
+     if (x_mean >= ci_min[i] && x_mean <= ci_max[i])
segments(i,ci_min[i],i,ci_max[i],col="gray") # a gray ci includes the true value
+     if (x_mean < ci_min[i] || x_mean > ci_max[i])
segments(i,ci_min[i],i,ci_max[i],col="black")    # a black ci does not include
                                                   the true value
+     }
> abline(h=x_mean,col="black",lw=2,lty="dashed")   # plot the true mean as a dashed line
> frac              # print the fraction of samples containing the true mean
[1] 0.92
```

3.6 Comparing the Means of Two Samples

Suppose you have two sets of measurements for a variable that were obtained on different days. When you calculate the mean of each sample, you will probably find that they are different and you may be interested in estimating whether the difference is significant in a *statistical* sense. The results of such a calculation would give you an indication as to whether the two samples were drawn from the same population or whether something happened that changed the population during the time period between the measurement sets.

You can use the *t* variable to compare the means of two samples to determine whether they are significantly different at a chosen confidence level *P*. For example, suppose we measure the lifetimes of birthday candles (in minutes). The data set x1 is a set of lifetimes measured during someone's birthday party and x2 is a set of lifetimes measured during another birthday party:

```
> x1 <- c(7.2,7.6,6.9,8.2,7.3,7.8,6.6,6.9,5.5,7.4,5.7,6.2)
> x2 <- c(7.5,8.7,7.7,7.5,6.7,11.2,7.0,10.7,7.0,8.6,6.1,6.3)
```

There is a built-in R function, t.test(), that can do this test:

```
> P = 0.95                        # confidence level
> t.test(x1,x2,conf.level=P)          # the t-test
        Welch Two Sample t-test
data:   x1 and x2
t = -1.8535, df = 16.289, p-value = 0.08201
alternative hypothesis: true difference in means is not equal to 0
95 percent confidence interval:
 -2.0885343   0.1385343
sample estimates:
mean of x mean of y
 6.941667   7.916667
```

What is relevant here is the *p*-value. A typical criterion to apply is the following: If the *p*-value is greater than $1 - P$ (= 0.05 here), then the sample means are not significantly different in a statistical sense. In the present case, the means differ by about 1 minute, but this difference is not statistically significant. A conclusion you would reach is that all of the birthday candles were likely to have come from the same population. This example actually corresponds to a hypothesis test, which is an interesting subject in itself.

3.7 Testing Data for Normality

Suppose we have a set of data and we want to assess whether these data are well described by a normal distribution. There is a built-in function, shapiro.test(), that can be used to test for normality:

```
> x <- rnorm(100, mean=10,sd=2)      # the data to be tested
> shapiro.test(x)

        Shapiro-Wilk normality test

data:   x
W = 0.98257, p-value = 0.2095
```

This *p*-value is large enough that the data can be assumed to be normal. Typically, it is assumed that the *p*-value should be no smaller than 0.05 for the assumption of normality to hold.

An alternative method for doing this is the chi-square (χ^2) test, which actually can be used to test whether data follow normal as well as other distributions, e.g., a uniform distribution. To use this test we would first generate a histogram for the data set, with the number of bins being some number K. We will define n_i to be the actual number of data points in bin i and n' to be the expected number of data points in bin i. The expected number of data points is $n'_i = P_i N$, where P_i is the probability that a data point would be in bin i and $N = \sum_{i=1}^{K} n_i$ is the total number of data points in the K bins that comprise the histogram. The variable χ^2 is calculated using

$$\chi^2 = \sum_{i=1}^{K} \frac{\left(n_i - n'_i\right)^2}{n'_i}. \tag{3.10}$$

The smaller χ^2 is, the better the fit is.

To proceed, we calculate the number of degrees of freedom,

$$y = K - 1 - r, \tag{3.11}$$

where r is the number of parameters that were employed, using the data, to calculate the bin probabilities P_i. For a normal distribution we have $r = 2$ because we use the data to calculate $\bar{x}$ and s_x. The level of significance, $\alpha = 1 - P$, corresponding to the calculated values of χ^2 and v is then determined. The value of α provides guidance as to how well the data are fit (described) by the probability density function (pdf). A fit is typically considered to be poor if roughly $\alpha < 0.05$, acceptable if $0.05 < \alpha < 0.95$, and exceptional if $\alpha > 0.95$.

Here is some R code to illustrate this procedure:

```
> x <- rnorm(300, mean=10,sd=2)       # the data to be tested
> N <- length(x)                      # number of data points
> x_mean <- mean(x)                   # data mean
> x_sd <- sd(x)                       # data standard deviation
> r <- 2                 # number of parameters estimated from the data
> x_hist <- hist(x,freq=FALSE)        # data histogram (prob)
> # standard normal curve
> curve(dnorm(x,mean=x_mean,sd=x_sd),add=TRUE,col="red",lw="2")
> z_breaks <- (x_hist$breaks - x_mean)/x_sd  # histogram break points
> n_prime <- numeric(length(z_breaks)-1)
> for (i in 1:length(n_prime)) {
> # the number of data points in a bin if the assumed pdf applies
+      n_prime[i] <- N*(pnorm(z_breaks[i+1]) - pnorm(z_breaks[i]))
+      }
> # calculate chi_squared
```

```
> chi_squared <- sum(((n_prime - x_hist$counts)^2)/n_prime)
> chi_squared                    # print chi-squared
[1] 9.119399
> dof <- length(n_prime)-1-r   # number of degrees of freedom
> dof                            # print the number of degrees of freedom
[1] 9
> alpha <- 1 - pchisq(chi_squared,df=dof)   # alpha value
> alpha                            # print alpha
[1] 0.4263265
```

The code yields $\alpha = 0.43$, indicating that the original data are normal. This was not unexpected because these data were drawn from a normal population. You should run this again, but using a uniform distribution to generate the original data set.

Another approach is to calculate the maximum allowable χ^2 value based on a critical level of significance you select and the number of degrees of freedom. This value (`chi_sq_max`) is simply calculated using the following R command:

```
> level_of_significance <- 0.05
> chi_sq_max <- qchisq(p=1-level_of_significance,df=dof)
```

If the value of χ^2 calculated using the experimental data is less than this maximum value, then it is assumed that the fit is acceptable. To use this method, though, you need to specify the critical level of significance, denoted in the code as `level_of_significance`. A critical level of significance value of 0.05 is typically used. It is also noted that the results you obtain will depend upon the bin limits for the histogram.

3.8 Outlier Identification

When you are doing experiments you are sometimes confronted with data points that somehow just seem wrong or do not appear to belong. Perhaps you made a mistake such as putting batteries in backward or flipping the wrong switch, in which case it is justifiable to reject such data. Conversely, you might not be able to identify any mistake that was made that would produce erroneous data. A question to be asked is then: *Should you keep all of the data, including data you think might be erroneous?* Of course, you should reject data you *know* are erroneous because of some mistake that was made, but what about the rest of the data? This is where the concept of outliers comes in. An outlier is defined here as

> *a data point that falls outside of the probable range of values.*

There are statistical tests for identifying outliers, for example, with box plots, as described previously, and with other methodologies, to be discussed in the following. Once outliers are identified, a decision needs to be made about whether or not to reject any of them.

Rejecting outliers is a controversial subject. One line of thought is that you should never reject data points unless you know for sure that they were contaminated by blunders. Another line of thought is to reject outliers on the assumption that they are statistically unlikely, so it is "probable" that some mistake caused the outliers to appear even if you do not know what the mistake was. Ultimately, though, it is up to you whether or not to reject an outlier. A reason you might want to reject outliers is that they may lead to overestimates of errors. In the following, we present two popular outlier detection methods: the modified Thompson τ technique and Chauvenet's criterion.

3.8.1 Modified Thompson τ Technique

This technique states that a data point can be considered to be an outlier if the criterion

$$\delta = |x_i - \bar{x}| \geq \tau s_x \qquad (3.12)$$

is satisfied. If it is not satisfied, then the data point is not an outlier. The variable τ is defined as

$$\tau = \frac{t_{v,1-\alpha/2}(N-1)}{\sqrt{N\left(N-2+t_{v,1-\alpha/2}^2\right)}}, \qquad (3.13)$$

where
   N is the number of data points
   $\alpha = 1 - P$, P is the confidence level
   t is Student's t variable
   $v = N - 2$ is the number of degrees of freedom to use with this test

A short presentation of τ as a function of N is given in Table 3.1 for $P = 0.95$.

This outlier test can be used multiple times, but the data mean and standard deviation should be recalculated every time an outlier is rejected before the test is readministered. Here is some code to illustrate the use of this test:

```
> x <- c(2.3,4.2,3.2,4.1,1.1,5.4,3.3,4.4,3.7,2.7,3.0,1.9,7.9,4.6,3.4)
> N <- length(x)              # number of data points
```

TABLE 3.1

The Variable τ as a Function of N for $P = 0.95$

N	3	5	10	20	50	100	500	1000
τ	1.151	1.571	1.798	1.885	1.931	1.946	1.957	1.959

For the modified Thompson τ technique.

```
> P <- 0.95                          # confidence level
> t <- qt(p=(1+P)/2,df=N-2)      # t value
> tau <- t*(N-1)/(sqrt(N)*sqrt(N-2+t^2))   # tau value
> deltaMax <- tau*sd(x)          # maximum allowable delta
> delta <- abs(x-mean(x))        # all delta values
> plot(delta,ylim=c(0,1.1*max(c(delta,deltaMax))))    # plot deltas
> abline(h=deltaMax,col="red")        # mark the maximum allowable delta
> which(delta >= deltaMax)     # identify potential outliers
[1] 13
```

The code identifies potential outliers, in this case data point 13, by using the which() function. Outliers also appear as points above the horizontal line in Figure 3.11. At this point, we could remove this data point with the command x <- x[-13] and then redo this test to check whether there are any more outliers.

3.8.2 Chauvenet's Criterion

Chauvenet's criterion states that a data point can be considered to be an outlier if it is outside a certain probability band around the data mean. In using

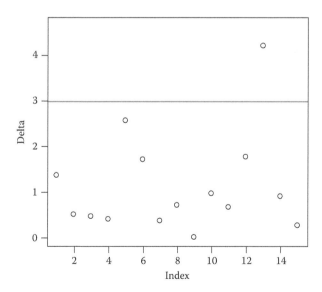

FIGURE 3.11
Plot of δ (delta) for each data point. Points above the horizontal line are potential outliers.

TABLE 3.2

Maximum z Values as a Function of N for Chauvenet's Criterion

N	3	5	10	20	50	100	300	500	1000
z_{max}	1.38	1.65	1.96	2.24	2.57	2.81	3.14	3.29	3.48

this criterion, we assume that the data are normally distributed and the criterion should only be used once; i.e., if outliers are rejected from a data set, we do not use Chauvenet's criterion to identify and reject any more outliers from the remaining data.

The Chauvenet probability band is defined as

$$P_{chauv} = 1 - \frac{1}{2N}, \qquad (3.14)$$

where N is the number of data points. Because we are assuming a normal distribution, we can calculate the magnitude of the z value, i.e., z_{max}, that corresponds to the band edges by using the qnorm() function. A short presentation of z_{max} vs. N is given in Table 3.2.

As an example, suppose that $N = 20$ such that $z_{max} = 2.24$. We would reject all data outside of the probability band, i.e., the shaded area in Figure 3.12. Here is the code:

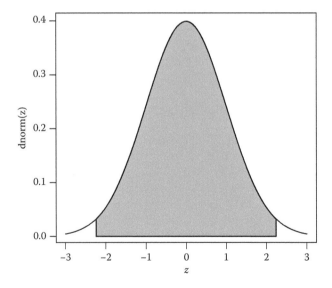

FIGURE 3.12
Chauvenet probability band for $N = 20$.

```
> z <- seq(from=-3,to=3,by=0.01)
> y <- dnorm(z)
> zmax <- 2.24
> plot(z,y,ylab="dnorm(z)",type="l")
> # shade the probability plot
> z_shade <- seq(from=-zmax,to=zmax,by=0.01)
> y_shade <- dnorm(z_shade)
> polygon(c(-zmax,z_shade,zmax),c(0,y_shade,0),
col=gray(0.7),border=NULL)
```

When z_i is calculated for a data point i with

$$z_i = \frac{x_i - \overline{x}}{s_x}, \tag{3.15}$$

we use the mean and standard deviation from the data set.

We illustrate the use of Chauvenet's criterion with the following example code:

```
> x <- c(2.3,4.2,3.2,4.1,1.1,5.4,3.3,4.4,3.7,2.7,3.0,1.9,7.9,4.6,3.4)
> N <- length(x)                      # number of data points
> z <- (x-mean(x))/sd(x)              # data z values
> p_chauv <- 1-1/(2*N)                # Chauvenet probability band value
> z_max <- qnorm(0.5+0.5*p_chauv)     # maximum allowable z value
> z_min <- -z_max                     # minimum allowable z value
> x_max <- mean(x) + z_max*sd(x)      # maximum allowable data value
> x_min <- mean(x) + z_min*sd(x)      # minimum allowable data value
> plot(x, ylim = c(min(x)-1,max(x)+1))    # plot the data
> abline(h=x_min,col="red",lw=2)      # mark minimum allowable data value
> abline(h=x_max,col="red",lw=2)      # mark maximum allowable data value
```

Note that in this code we calculate the maximum and minimum allowable values of x based on the allowable z values:

$$x_{min} = \overline{x} - s_x z_{max}, \tag{3.16}$$

$$x_{max} = \overline{x} + s_x z_{max}. \tag{3.17}$$

According to Chauvenet's criterion, data points that do not lie between the two horizontal lines in Figure 3.13 may be outliers. In this case, there is only one potential outlier, i.e., data point 13. The mean and standard deviation of the data set are 3.68 and 1.61, respectively, when data point 13 is included, but are reduced to 3.38 and 1.15, respectively, when it is not. Throwing out the outlier makes a difference, but in the end, it is up to you to determine whether you would keep this outlier or reject it.

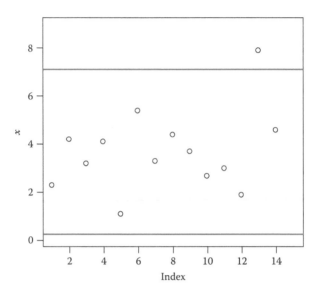

FIGURE 3.13
Chauvenet probability band limits (horizontal lines) and data points (circles).

Problems

P3.1   A sample x is defined as follows:

```
x <- c(2.3,4.2,3.2,4.1,1.1,5.4,3.3,4.4,3.7,2.7)
```

Write your own code to directly calculate the following quantities: mean, variance, standard deviation, and median. Compare your results with results from internal R functions.

P3.2   The variables x and y depend on the time t as $x = \sin(t)$ and $y = \cos(t)$. Plot x and y as functions of t on the same graph for $-\pi < t < \pi$. After creating vectors for x and y with $N = 1 \times 10^6$ elements each, plot y as a function of x and calculate the correlation coefficient ρ_{xy}. Is it what you would expect? Explain.

P3.3   Two samples x and y are defined as follows:

```
x <- c(3.0,2.9,2.7,2.7,2.9,2.7,3.6,2.7,2.2,3.5)
y <- c(2.4,3.2,2.7,3.3,3.0,3.1,3.6,2.0,3.7,2.1)
```

Write your own code to directly calculate the correlation coefficient ρ_{xy}. Compare your results with results from internal R functions. What happens to ρ_{xy} if both x and y are sorted from smallest to largest?

P3.4 Benford's law is an empirical expression that can be used to estimate the probability that a certain digit will appear first in a large collection of numbers that describe natural phenomena. It can also be used to detect fraud, i.e., data fabrication.

Let D_1 be the first significant (nonzero) digit of a number. For the number 0.00314, $D_1 = 3$, and for the number 54.334, $D_1 = 5$. Benford's law states that the probability P that $D_1 = d$, where $d \in \{1,2,3,4,5,6,7,8,9\}$, is given by $P(D_1 = d) = \log_{10}[(d + 1)/d]$. For example, $P(D_1 = 2) = \log_{10}[(2 + 1)/2] = 0.1761$.

Using Benford's law with the chi-square goodness-of-fit test, evaluate whether the following numbers, which are from a report, are likely fabricated or whether they could represent natural variations in a physical quantity:

0.307	0.557	0.873	0.501	0.923	0.381	0.069	0.343	0.336	0.868
0.703	0.400	0.961	0.667	0.109	0.090	0.766	0.262	0.691	0.687
0.200	0.956	0.344	0.099	0.489	0.975	0.889	0.740	0.342	0.454
0.527	0.981	0.230	0.608	0.149	0.347	0.086	0.222	0.484	0.179
0.069	0.127	0.006	0.663	0.734	0.355	0.580	0.318	0.496	0.036
0.530	0.820	0.265	0.435	0.556	0.657	0.833	0.517	0.971	0.338
0.076	0.409	0.271	0.129	0.446	0.031	0.652	0.684	0.576	0.073
0.863	0.998	0.873	0.623	0.612	0.952	0.819	0.240	0.366	0.885
0.223	0.092	0.764	0.753	0.955	0.567	0.597	0.861	0.434	0.950
0.309	0.808	0.652	0.335	0.836	0.222	0.629	0.358	0.943	0.241
0.355	0.209	0.035	0.194	0.584	0.828	0.494	0.086	0.249	0.125
0.635	0.891	0.091	0.515	0.398	0.776	0.005	0.107	0.186	0.234
0.265	0.211	0.876	0.647	0.775	0.838	0.697	0.959	0.081	0.681
0.415	0.360	0.720	0.563	0.537	0.257	0.574	0.635	0.957	0.780
0.157	0.980	0.654	0.588	0.220	0.907	0.341	0.637	0.212	0.932
0.663	0.308	0.821	0.012	0.016	0.969	0.982	0.083	0.090	0.898
0.241	0.788	0.459	0.216	0.115	0.744	0.054	0.526	0.905	0.594
0.337	0.587	0.825	0.816	0.799	0.366	0.267	0.068	0.482	0.862
0.950	0.012	0.711	0.722	0.561	0.259	0.518	0.438	0.239	0.597
0.348	0.038	0.171	0.264	0.325	0.141	0.025	0.203	0.242	0.786

P3.5 Write your own code to calculate the 95% confidence interval for the population mean using the following sample:

```
x <- c(2.3,4.2,3.2,4.1,1.1,5.4,3.3,4.4,3.7,2.7,3.0,1.9,7.9,4.6,
   3.4)
```

Compare your results with the R function t.test(). How would your results change if this set of numbers was actually the population?

P3.6   Write your own code to calculate the 95% confidence interval for the population variance using the following sample:

```
x <- c(2.3,4.2,3.2,4.1,1.1,5.4,3.3,4.4,3.7,2.7,3.0,1.9,7.9,4.6,
3.4)
```

Compare your results with the R function `sigma.test()`. How would your results change if this set of numbers was actually the population?

P3.7   A data sample has the following four values:

```
x <- c(1,2,3,1000)
```

Evaluate whether any of these values would be flagged as outliers by applying Chauvenet's criterion as well as the modified Thompson τ technique.

P3.8   Create a data sample with the following R code:

```
set.seed(100)
x <- rnorm(n=20,mean=10,sd=1)
```

Evaluate whether any of these values would be flagged as outliers by applying Chauvenet's criterion as well as the modified Thompson τ technique. If so, do the mean and standard deviation change significantly if outliers are rejected?

P3.9   Run the code that created Figure 3.10 but for various values of n and P and for normal and uniform populations. Describe how the confidence intervals are influenced by n, P, and the population distribution.

P3.10 The central limit theorem states that the distribution of mean values of repeated samples of size N from a large set of random numbers should be approximately normal. Evaluate this statement using R code and for random numbers from uniform and normal populations. Investigate various values of N.

References

1. J. Verzani, *Using R for Introductory Statistics*, 2nd edn., CRC Press, New York, 2014.
2. N. Radziwill, *Statistics (The Easier Way) with R: An Informal Text on Applied Statistics*, Lapis Lucera, San Francisco, CA, 2015.
3. M. J. Crawley, *The R Book*, 2nd edn., John Wiley & Sons, West Sussex, U.K., 2013.
4. T. Hothorn and B. S. Everitt, *A Handbook of Statistical Analyses using R*, 3rd edn., CRC Press, Boca Raton, FL, 2014.

4

Curve Fits

After conducting experiments we often want to fit some type of equation to the data. In this chapter, we will discuss a few different approaches, namely, linear regression, nonlinear regression, and kernel smoothing. It is also possible to generate confidence intervals for curve fits, but we will delay the discussion of this until a later chapter. Some useful references on curve fitting are [1–7].

4.1 Linear Regression

Linear regression is a commonly used curve-fitting technique. The word "linear" refers to how coefficients appear in the mathematical equation that is being used for the fit (i.e., the coefficients all appear linearly). More precisely, a linear regression function is composed of a set of functions that are each multiplied by a curve-fit coefficient and then added together. For example, if we have a bivariate data set x and y, we could work with an expression of the form

$$y_c = a_0 f_0(x) + a_1 f_1(x) + a_2 f_2(x) + \cdots + a_n f_n(x), \qquad (4.1)$$

where $f_0(x), f_1(x), f_2(x),\ldots$ could be nonlinear functions of x and the subscript c on y_c indicates a curve fit. However, the coefficients $a_0, a_1,\ldots$ appear linearly so it would be a linear regression problem to determine them.

We define the residual E_i as the difference between data point y_i and the predicted value at x_i:

$$E_i = y_i - y_c(x_i). \qquad (4.2)$$

We then combine the residuals in some fashion into an objective function E. This objective function is actually a function of the regression coefficients, i.e., $E = E(a_0, a_1, a_2,\ldots, a_n)$, and the goal is to find the regression coefficients that make the magnitude of E as small as possible.

With linear regression, E is typically defined to be the sum of the squares of the residuals:

$$E = \sum_{i=1}^{N} E_i^2. \tag{4.3}$$

This is convenient because it can allow the optimal coefficients to be determined analytically. It is possible, however, with the use of computers to employ other definitions of E and then minimize E using some search process. For example, E could be defined as shown in the following equations:

$$E = \sum_{i=1}^{N} |E_i|, \tag{4.4}$$

$$E = \left| \sum_{i=1}^{N} E_i \right|, \tag{4.5}$$

$$E = \sum_{i=1}^{N} E_i^4. \tag{4.6}$$

When E is as defined in Equation 4.3, it is possible to solve for the coefficients $a_0, a_1,\ldots$ using basic linear algebra and calculus. By setting $\partial E / \partial a_0 = \partial E / \partial a_1 = \cdots = \partial E / \partial a_n = 0$, we can derive

$$A = \left(F^{\mathsf{T}} F\right)^{-1} F^{\mathsf{T}} Y, \tag{4.7}$$

where

$$A = \begin{bmatrix} a_0 \\ a_1 \\ \vdots \\ a_n \end{bmatrix},$$

$$Y = \begin{bmatrix} y_1 \\ y_2 \\ \vdots \\ y_N \end{bmatrix},$$

$$F = \begin{bmatrix} f_0(x_1) & f_1(x_1) & f_2(x_1) & \cdots & f_n(x_1) \\ f_0(x_2) & f_1(x_2) & f_2(x_2) & \cdots & f_n(x_2) \\ \vdots & & \vdots & & \vdots \\ f_0(x_N) & f_1(x_N) & f_2(x_N) & \cdots & f_n(x_N) \end{bmatrix},$$

and the superscript "T" denotes transpose.

In R, Equation 4.7 (or an equivalent) is implemented with the function `lm()`. As an example, we will fit the function

$$y_c = a_0 + a_1 x \tag{4.8}$$

to the data $x = \{1, 2, 3, 4, 5\}$, $y = \{0.95, 1.51, 1.99, 2.59, 3.01\}$. The following R code determines the coefficients:

```
> x <- c(1,2,3,4,5)
> y <- c(0.95,1.51,1.99,2.59,3.01)
> model <- lm(y~x)
> coefficients(model)
(Intercept)          x
       0.45       0.52
```

We thus find that $a_0 = 0.45$ and $a_1 = 0.52$.

The command `lm(y~x)` solves the problem and the result is stored in `model`. The "~" tells R that y is to be considered a function of x. When `lm()` is used, we specify only the functions $f_1(x), f_2(x)$, etc.; e.g., $f_1(x) = x$ here. The coefficients $a_0, a_1, \ldots$ are not entered explicitly. The function `I()` can be used if you want to enter a specific functional relationship; e.g., use `lm(y~I(exp(x)))` if you want to have $y_c = a_0 + a_1 e^x$. If you want to explicitly set $a_0 = 0$, use `y~0+` in the `lm()` function; e.g., `lm(y~0+I(exp(x)))` corresponds to $y_c = a_1 e^x$. The command `coefficients()` gives the coefficients for the fit.

We can also approach this problem in a more direct fashion by calculating E for different values of the coefficients a_0 and a_1, and then finding the coefficient values that minimize E. This is accomplished with the following R code:

```
> library(fields)
> library(plot3D)
> x <- c(1,2,3,4,5)
> y <- c(0.95,1.51,1.99,2.59,3.01)
> N <- length(x)
> a0 <- seq(from=0,to=1,length.out=500)
> a1 <- seq(from=0,to=1,length.out=500)
> E <- matrix(nrow=length(a0),ncol=length(a1))
> for (i in 1:length(a0)) {
+   for (j in 1:length(a1)) {
+       yc <- a0[i]+a1[j]*x
```

```
+              eps <- yc-y
+              E[i,j] <- sum(eps^2)
+ } }
> image.plot(x=a0,y=a1,z=log10(E))
> contour(x=a0,y=a1,z=log10(E),add=TRUE)
> vals <- which(E == min(E),arr.ind=TRUE)
> a0[vals[1]]
[1] 0.4529058
> a1[vals[2]]
[1] 0.5190381
> min(E)
[1] 0.006409255
> quartz()
> M <- mesh(a0,a1)
> surf3D(M$x,M$y,log10(E),bty='b2',ticktype='detailed',
+ xlab='a0',ylab='a1',zlab='log10(E)',phi=0,theta=45)
```

This code calculates E for discrete values of a_0 and a_1 and plots the results in two different ways. Figure 4.1a shows a contour plot of $\log_{10}(E)$ in the a_0–a_1 plane, and Figure 4.1b shows a surface plot of $\log_{10}(E)$ as a function of a_0 and a_1. Logarithms are used here because they vary strongly for small values of E. The values for the coefficients, i.e., where E is minimized, are found to be $a_0 = 0.45$ and $a_1 = 0.52$, in agreement with lm().

In some cases you will not know the order of the polynomial to be fit, but there are statistical approaches that can be used to determine what is appropriate (at least in a statistical sense). We illustrate this with an example in which we add some noise to the function $y = 1 + 2x + 3x^2$ for the range $0 \leq x \leq 1$. We will then use lm() to try to recover the coefficient of the original function by fitting

$$y_c = a_0 + a_1 x + a_2 x^2 + \cdots + a_n x^n \tag{4.9}$$

to the noisy data. Our goal is to determine the lowest-order polynomial with coefficients that are statistically significant. To do this in R is straightforward, but we need to make sure that we use the correct syntax.

Here is some R code to address this problem. We begin with a simple model of the form $y = a_1 x$:

```
> x <- c(0,0.1,0.2,0.3,0.4,0.5,0.6,0.7,0.8,0.9,1)
> yExact <- 1+2*x+3*x^2
> set.seed(100)
> y <- yExact + 0.25*rnorm(length(x))   # the noisy data
> Model_0 <- lm(y~0+I(x))         # y = a1*x
> summary(Model_0)

Call:
lm(formula = y ~ 0 + I(x))
```

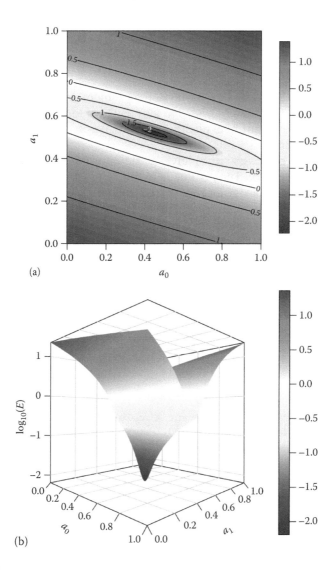

(a)

(b)

FIGURE 4.1
(a) Contour plot showing $\log_{10}(E)$ as a function of a_0 and a_1. (b) Surface plot showing $\log_{10}(E)$ as a function of a_0 and a_1.

```
Residuals:
     Min        1Q    Median        3Q       Max
-0.32622  -0.05272   0.01107   0.35400   0.87445

Coefficients:
     Estimate Std. Error t value Pr(>|t|)
I(x)    5.7679      0.2134    27.03 1.11e-10 ***
---
Signif. codes:  0 '***' 0.001 '**' 0.01 '*' 0.05 '.' 0.1 ' ' 1
```

```
Residual standard error: 0.4187 on 10 degrees of freedom
Multiple R-squared:  0.9865,     Adjusted R-squared:  0.9851
F-statistic: 730.5 on 1 and 10 DF,  p-value: 1.112e-10
```

The command `lm(y~0+I(x))` solves the problem and the result is stored in `Model_0`. The function `I()` is what you use to enter a specific functional relationship. If you want to explicitly set $a_0 = 0$, use `y~0+` in the `lm()` function, as we did earlier.

The command `summary(ans)` gives basic information about the fit. The coefficient is $a_1 = 5.7679$. The p-value is in the `Pr(>|t|)` column, and if we use a threshold level of significance of 0.05, then the a_1 term in the proposed equation is statistically significant. However, we will not stop here. In the following, we consider models of different complexity:

```
> Model_1 <- lm(y~I(x))            # y = a0+a1*x
> summary(Model_1)

Call:
lm(formula = y ~ I(x))

Residuals:
    Min       1Q    Median       3Q      Max
-0.40821 -0.21849  0.01107  0.15350  0.50050

Coefficients:
             Estimate Std. Error t value Pr(>|t|)
(Intercept)    0.5740     0.1593   3.602  0.00573 **
I(x)           4.9480     0.2693  18.372 1.92e-08 ***
    ---
Signif. codes:  0 '***' 0.001 '**' 0.01 '*' 0.05 '.' 0.1 ' ' 1

Residual standard error: 0.2825 on 9 degrees of freedom
Multiple R-squared:  0.974,  Adjusted R-squared:  0.9711
F-statistic: 337.5 on 1 and 9 DF,  p-value: 1.917e-08

> Model_2 <- lm(y~I(x)+I(x^2))       # y = a0+a1*x+a2*x^2
> summary(Model_2)

Call:
lm(formula = y ~ I(x) + I(x^2))

Residuals:
    Min       1Q    Median       3Q      Max
-0.19313 -0.06163 -0.01394  0.07901  0.18666

Coefficients:
             Estimate Std. Error t value Pr(>|t|)
(Intercept)    0.9573     0.1070   8.946 1.94e-05 ***
```

```
I(x)                2.3927      0.4979   4.806 0.001345 **
I(x^2)              2.5553      0.4795   5.329 0.000703 ***
---
Signif. codes:  0 '***' 0.001 '**' 0.01 '*' 0.05 '.' 0.1 ' ' 1

Residual standard error: 0.1405 on 8 degrees of freedom
Multiple R-squared:  0.9943,  Adjusted R-squared:  0.9929
F-statistic: 696.7 on 2 and 8 DF,  p-value: 1.062e-09

> Model_3 <- lm(y~I(x)+I(x^2)+I(x^3))   # y = a0+a1*x+a2*x^2+a3*x^3
> summary(Model_3)

Call:
lm(formula = y ~ I(x) + I(x^2) + I(x^3))

Residuals:
     Min        1Q     Median        3Q       Max
-0.136725 -0.079201   0.006139  0.032746  0.232577

Coefficients:
            Estimate Std. Error t value Pr(>|t|)
(Intercept)   0.8683     0.1164   7.457 0.000142 ***
I(x)          3.8059     1.0604   3.589 0.008868 **
I(x^2)       -1.1506     2.5397  -0.453 0.664228
I(x^3)        2.4706     1.6667   1.482 0.181804
---
Signif. codes:  0 '***' 0.001 '**' 0.01 '*' 0.05 '.' 0.1 ' ' 1

Residual standard error: 0.131 on 7 degrees of freedom
Multiple R-squared:  0.9957,      Adjusted R-squared:  0.9938
F-statistic: 534.8 on 3 and 7 DF,  p-value: 1.256e-08
```

As we add powers of x, we find that all terms are statistically significant if the highest power of x is x^2, but if we try a third-order polynomial, then the x^3 term is not statistically significant. As a result, we stop with the second-order polynomial from Model_2:

$$y = 0.9573 + 2.3927x + 2.5553x^2.$$

We plot this fit (dashed line) along with the exact (black line) and noisy data (circles) in Figure 4.2, where it is evident that the curve fit does an excellent job of reproducing the original noise-free equation. Here is the code:

```
> xPlot <- seq(from=0,to=1,by=0.01)
> yExactPlot <- 1+2*xPlot+3*xPlot^2
> plot(xPlot,yExactPlot,xlab="x",ylab="y",type="l")
```

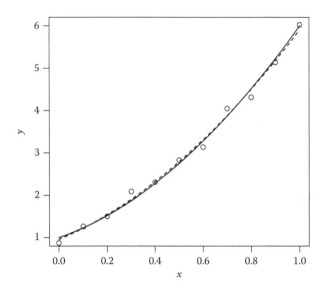

FIGURE 4.2
Polynomial fit (dashed line) along with the exact fit (black line) and data (circles).

```
> yFitPlot <- 0.9573+2.3927*xPlot+2.5553*xPlot^2
> lines(xPlot,yFitPlot,col="black",Hy="dashed")
> points(x,y)
```

Note that, with these types of analyses, at least with polynomials, you should keep increasing the order of the polynomials until statistically insignificant terms appear. Then you can use the highest-order polynomial where the highest-order term is statistically significant. Unless you have good reason otherwise, e.g., based on the physics, it has been suggested [5] that you should retain all lower-order terms regardless of their p-values. Of course, if the physics of your problem dictates that you should use a specific type of polynomial, then this is what you should use.

4.2 Nonlinear Regression

It is sometimes the case that you need to fit a curve to data but where at least one of the curve-fit parameters appears nonlinearly. Here are some examples:

$$y_c = a_0 x^{a_1},$$
$$y_c = a_0 + x + \ln(1 + a_1 x),$$
$$y_c = a_0 x \sin(a_1 x).$$

In some cases, the curve-fit equation can be transformed into one that is linear in terms of the parameters, but if this is not possible, we can use the `nls()` function in R for the curve fitting.

As an example, we will fit an equation of the form $y_c = a_0 x^{a_1}$ to a set of data. This equation can actually be made linear in the coefficients by taking logarithms, but we will not do this here. Here is the code:

```
> x <- c(32,50,100,150,212)                  # x data
> y <- c(0.337,0.345,0.365,0.380,0.395)      # y data
> plot(x,y)                                  # plot the data points
> nlmod <- nls(y~a0*(x^a1),start=list(a0=0.1,a1=0.1))    # curve fit
> summary(nlmod)                             # data fit summary


Formula: y ~ a0 * (x^a1)


Parameters:
   Estimate Std. Error t value Pr(>|t|)
a0 0.248780   0.006473   38.44 3.87e-05 ***
a1 0.084996   0.005654   15.03 0.000639 ***
---


Signif. codes:  0 '***' 0.001 '**' 0.01 '*' 0.05 '.' 0.1 ' ' 1


Residual standard error: 0.003155 on 3 degrees of freedom


Number of iterations to convergence: 4
Achieved convergence tolerance: 7.928e-06
```

The curve-fit parameters are thus $a_0 = 0.25$ and $a_1 = 0.085$. Both are statistically significant. We can also generate a plot of the original data and the curve fit (Figure 4.3) from

```
> xFit <- seq(from=min(x),to=max(x),by=(max(x)-min(x))/10)
> yFit <- 0.248780*xFit^0.084996    # predicted y values
> lines(xFit,yFit,col="black")
```

Figure 4.3 shows the data points (circles) as well as the fitted line. Note that when `nls()` is used, you need to provide initial estimates for the curve-fit parameters. In the present case, this was done with the `start=list(a0=0.1,a1=0.1)` command.

Similar to what was done with the earlier linear regression problem, we can also treat a nonlinear regression problem as an optimization problem where we minimize an objective function. To this end, we will use Equations 4.2 and 4.3 with $y_c = a_0 x^{a_1}$. The R code is as follows:

```
> library(fields)
> library(plot3D)
> x <- c(32,50,100,150,212)                              # x data
```

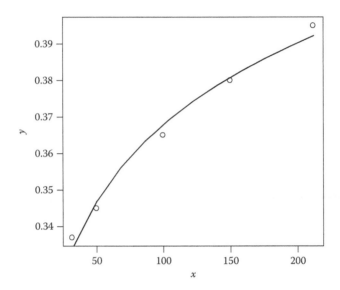

FIGURE 4.3
Nonlinear curve fit (black line) and the original data (circles).

```
> y <- c(0.337,0.345,0.365,0.380,0.395)    # y data
> a0 <- seq(from=0,to=0.5,length.out=500)
> a1 <- seq(from=0,to=0.2,length.out=500)
> E <- matrix(nrow=length(a0),ncol=length(a1))
> for (i in 1:length(a0)) {
+     for (j in 1:length(a1)) {
+             yc <- a0[i]*(x^a1[j])
+             eps <- y-yc
+             E[i,j] <- sum(eps^2)
+     }
+ }
> vals <- which(E == min(E),arr.ind=TRUE)
> a0[vals[1]]
[1] 0.248497
> a1[vals[2]]
[1] 0.08537074
> min(E)
[1] 3.011746e-05
> image.plot(x=a0,y=a1,z=log10(E))
> contour(x=a0,y=a1,z=log10(E),add=TRUE)
> quartz()
> M <- mesh(a0,a1)
> surf3D(M$x,M$y,log10(E),bty='b2',ticktype='detailed',
+ xlab='a0',ylab='a1',zlab='log10(E)',phi=0,theta=45)
```

This indicates that $a_0 = 0.25$ and $a_1 = 0.085$. The corresponding contour and surface plots for E as a function of a_0 and a_1 are shown in Figure 4.4a and b, respectively.

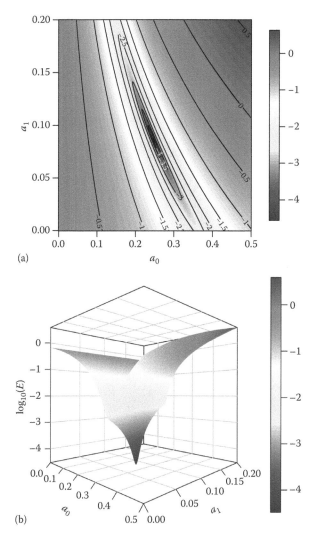

FIGURE 4.4
(a) Contour plot showing $\log_{10}(E)$ as a function of a_0 and a_1. (b) Surface plot showing $\log_{10}(E)$ as a function of a_0 and a_1.

4.3 Kernel Smoothing

Kernel smoothing methods [6] can be useful in cases where polynomials or other functions may not be appropriate. A kernel smoothing method estimates the fit at a specific point, but nearby data points are weighted more strongly than data points further away. Such a fit is said to be "data driven" in that it is not tied to any particular functional form.

Here, we consider the method of local polynomial fitting [7]. In the local polynomial method, one assumes that the curve fit can be written as

$$y_c(x) = \sum_{j=0}^{n} a_j (x - x_0)^j = a_0 + a_1 (x - x_0) + \cdots + a_n (x - x_0)^n \qquad (4.10)$$

for x near x_0. It can be easily shown that a_1 is the first derivative of y_c at x_0, and a_0 is the value of y_c at x_0.

We minimize an objective function (E) in

$$E = \sum_{i=1}^{N} (y_i - y_c(x_i))^2 K_i \qquad (4.11)$$

by varying the coefficients a_j. Here, N is the total number of data points, x_0 is the particular value where the fit is being applied, n is the order of the polynomial, K_i is the kernel, and h is the bandwidth. We use a Gaussian kernel $K_i = \exp[-(x_i - x_0)^2/(2h^2)]$, though other kernels can be employed. The bandwidth h is an adjustable parameter that essentially allows for multiresolution analysis. Small values of h will cause the fit to more closely follow fluctuations in the data, while increasing h will smooth the fit by filtering out shorter timescale fluctuations in the data. We need to minimize Equation 4.11 at every x_0 value in which we are interested.

The simplest local polynomial is $y_c(x_0) = a_0$. For this case, minimizing Equation 4.11 yields

$$a_0 = \frac{\sum_{i=1}^{N} y_i K_i}{\sum_{i=1}^{N} K_i}. \qquad (4.12)$$

If we instead use $y_c(x) = a_0 + a_1(x - x_0)$, we will obtain

$$a_0 = \frac{\left(\sum_{i=1}^{N} y_i K_i\right)\left(\sum_{i=1}^{N}(x_i - x_0)^2 K_i\right) - \left(\sum_{i=1}^{N}(x_i - x_0) y_i K_i\right)\left(\sum_{i=1}^{N}(x_i - x_0) K_i\right)}{\left(\sum_{i=1}^{N} K_i\right)\left(\sum_{i=1}^{N}(x_i - x_0)^2 K_i\right) - \left(\sum_{i=1}^{N}(x_i - x_0) K_i\right)^2}, \qquad (4.13)$$

$$a_1 = \frac{\left(\sum_{i=1}^{N} K_i\right)\left(\sum_{i=1}^{N}(x_i - x_0) y_i K_i\right) - \left(\sum_{i=1}^{N} y_i K_i\right)\left(\sum_{i=1}^{N}(x_i - x_0) K_i\right)}{\left(\sum_{i=1}^{N} K_i\right)\left(\sum_{i=1}^{N}(x_i - x_0)^2 K_i\right) - \left(\sum_{i=1}^{N}(x_i - x_0) K_i\right)^2}. \qquad (4.14)$$

The R function `locpol()` provides an implementation of the local polynomial method. We will use this function for a fit to data composed of a sine wave with normal noise added. We will also extract estimates of the slope dy/dx and compare this with the derivative of the noise-free data. Here is the R code:

```
> library(locpol)
>
> # set basic parameters
> n <- 301
> x_min <- 0
> x_max <- 10
> d_x <- ((x_max - x_min)/(n-1))
> x <- seq(x_min,x_max,d_x)
>
> # define the test function y(x) and its derivative
> y <- sin(x)
> dydx <- cos(x)
>
> # add normal noise
> set.seed(100)
> y_noisy <- y + 0.1*rnorm(length(y))
>
> # calculate the fit
> bw <- 0.3
> deg <- 2
> data <- data.frame(y_noisy, x)
> fit <- locpol(y_noisy ~ x, data=data, deg=deg, xeval=x, kernel=gaussK, bw=bw)
>
> # save relevant model output
> bw <- fit$bw                      # bandwidth used for fitting
> y_fit <- fit$lpFit$y_noisy        # fitted y values
> dydx_fit <- fit$lpFit$y_noisy1    # fitted dy/dx values
>
> # plot the fit and the noisy data
> plot(x,y_noisy,type="p",pch=".",cex=4,xlab="x",ylab="y")
> lines(x,y_fit,type="l",col="black",lwd=2)
>
> # plot the fit and the original noise-free function
> quartz()
> plot(x,y_fit,type="l",col="black",lwd=2)
> lines(x,y,col="black",lwd=2,Hy="dashed")
>
> # plot dy/dx for the fit and the noise-free function
> quartz()
> plot(x,dydx_fit,type="l",col="blue",xlab="x",ylab="dy/dx",lwd=2)
> lines(x,dydx,col="black",lwd=2,Hy="dashed")
```

Figure 4.5 shows a plot of the noisy data (black points) along with the curve that was obtained from the fit by `locpol()`.

Figures 4.6 and 4.7 show comparisons between the noise-free y and dy/dx profiles (the solid lines) as well as the results from the `locpol()` fit (the dashed line). It is difficult to distinguish between the curves for y, indicating that the method worked very well. For dy/dx, however, there are some obvious differences especially near the beginning and ending x values, but overall the local polynomial method performed well for estimating slopes. The performance of the local polynomial method depends on the bandwidth and the order of the polynomial, so it is necessary to vary these parameters to obtain reasonable fits.

Uncertainty Analysis of Experimental Data with R

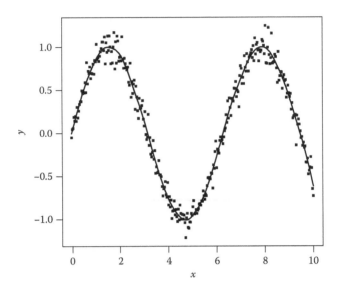

FIGURE 4.5
Plot of noisy data (black points) and a line fit.

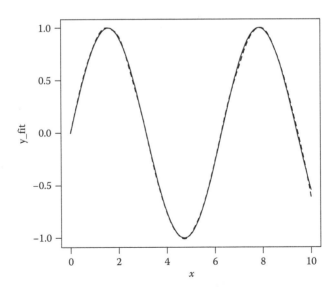

FIGURE 4.6
Comparison of noise-free *y* profile (solid lines) and the results from the `locpol()` fit (dashed line).

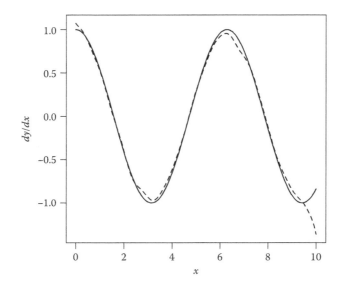

FIGURE 4.7
Comparison of noise-free dy/dx profile (solid line) and the results from the `locpol()` fit (dashed line).

Problems

P4.1 Consider the following data:

x	y
3	1.38
4	1.54
5	1.65
6	1.73
7	1.80
8	1.87
9	1.91
10	1.96
15	2.13
20	2.24
25	2.33
50	2.57
100	2.81
300	3.14
500	3.29
1000	3.48

Assume that a curve fit can be written as follows:

$$y_c = a_0 + a_1 w + a_2 w^2 + \cdots + a_n w^n,$$

where $w = \ln(x)$. Use the function lm() in R to find curve fits for $n = \{1, 2, 3, 4, 5, 6, 7\}$. What is the highest-order polynomial that is statistically significant?

P4.2 Starting with Equation 4.7, derive expressions for a_0 and a_1 for the curve fit $y_c = a_0 + a_1 x$.

P4.3 Fit an equation of the form $y_c = a_0 + a_1 x$ to the following data:

```
x <- c(1,2,3,4,5)
y <- c(0.98,1.55,1.89,2.49,3.31)
```

That is, find numerical values for a_0 and a_1. Do this by applying the objective functions in Equations 4.3 through 4.6 as well as by modifying the R code that produced Figures 4.1a and b. Are your results appreciably influenced by the choice of objective function?

P4.4 Fit an equation of the form $y_c = a_0 x^{a_1}$ to the following data:

```
x <- c(32,50,110,150,212)
y <- c(0.337,0.345,0.365,0.380,0.395)
```

Do this by expressing this equation in a linear regression form $Y_c = A_0 + A_1 X$ by taking logarithms and find numerical values for a_0, a_1, A_0, and A_1.

P4.5 Fit an equation of the form $y_c = a_0 x^{a_1}$ to the following data:

```
x <- c(1,2,3,4,5)
y <- c(0.98,1.55,1.89,2.49,3.31)
```

Do this by applying the objective functions in Equations 4.3 through 4.6 as well as by modifying the R code that produced Figures 4.4a and b. Are your results appreciably influenced by the choice of objective function?

P4.6 Derive Equation 4.12.

P4.7 Derive Equations 4.13 and 4.14.

P4.8 Apply Equation 4.12 to the data generated by the following R code using a Gaussian kernel:

```
> N <- 50
> x <- seq(from=0,to=4*pi,length.out=N)
> set.seed(100)
> y <- exp(-x/5)*sin(x)+rnorm(n=N,mean=0,sd=0.1)
```

Also generate a curve fit using `locpol()` with a Gaussian kernel. Plot your results from both curve fits with the data for various bandwidths. Comment on differences between the solutions.

P4.9 Apply Equations 4.13 and 4.14 to the data generated by the following R code using a Gaussian kernel:

```
> N <- 50
> x <- seq(from=0,to=4*pi,length.out=N)
> set.seed(100)
> y <- exp(-x/5)*sin(x)+rnorm(n=N,mean=0,sd=0.1)
```

Also generate curve fits for y_c and its first derivative using `locpol()` with a Gaussian kernel. Plot your results for curve fits with the data for various bandwidths. Comment on differences between the solutions.

P4.10 Evaluate Equations 4.12 and 4.13 for the limit where the bandwidth approaches zero and also for the bandwidth approaching infinity. Use a Gaussian kernel. What do plots of curve fits look like for these limits for the data generated by the following R code?

```
> N <- 20
> x <- seq(from=0,to=10,length.out=N)
> set.seed(100)
> y <- x*(10-x)+rnorm(n=N,mean=0,sd=4)
```

References

1. J. Verzani, *Using R for Introductory Statistics*, 2nd edn., CRC Press, New York, 2014.
2. N. Radziwill, *Statistics (The Easier Way) with R: An Informal Text on Applied Statistics*, Lapis Lucera, San Francisco, CA, 2015.
3. M. J. Crawley, *The R Book*, 2nd edn., John Wiley & Sons, West Sussex, U.K., 2013.
4. T. Hothorn and B. S. Everitt, *A Handbook of Statistical Analyses Using R*, 3rd edn., CRC Press, Boca Raton, FL, 2014.
5. J. J. Faraway, *Practical Regression and ANOVA Using R*, J.J. Faraway, Bath, U.K., 2002.
6. M. P. Wand and M. C. Jones, *Kernel Smoothing*, Chapman & Hall/CRC, New York, 1995.
7. J. Fan and I. Gijbels, *Local Polynomial Modeling and Its Applications*, Chapman & Hall/CRC, New York, 1996.

5

Uncertainty of a Measured Quantity

5.1 What Is Uncertainty?

Measurements of continuous quantities are reasonably assumed to never be exactly correct; i.e., there will always be errors (and even in the unlikely event that a measurement result was exactly correct, how would you know that this was the case). We generally want to be able to say something about how large or small these errors are and this is where uncertainties come into the picture [1–5]. As mentioned in [3], uncertainty can be considered to be a nonnegative parameter characterizing the dispersion of values attributed to a measurand (the quantity being measured). We will use the *standard deviation* for this purpose. To proceed forward and to allow for derivation of important relations, we next introduce concepts related to random variables.

5.2 Random Variables

We will now consider random variables. For the purposes of this textbook, we consider a random variable to be a numerical quantity whose value can be predicted only with a certain probability. Any quantity that is measured can be treated as a random variable. For continuous quantities, we really need to consider the probability that a measured value will fall within a specified range, which brings in probability density functions (pdfs; which are discussed in some detail in the Appendix).

Let $p(x)$ be a pdf for a single random variable x and $g(x)$ be a function of x. The expectation $\langle g(x) \rangle$, which is an average of $g(x)$, is given by

$$\langle g(x) \rangle = \int_{x_{min}}^{x_{max}} g(x) p(x) dx. \tag{5.1}$$

The expectation of $g(x) = x$ is the population mean, i.e., $\langle x \rangle = \mu_x$, which is given by

$$\mu_x = \int_{x_{min}}^{x_{max}} xp(x)dx. \tag{5.2}$$

It can also be shown that $\langle Ax \rangle = A\mu_x$ and $\langle A + x \rangle = A + \mu_x$ if A is a constant. The population variance, which we denote as $V(x)$, is the expectation of $(x - \mu_x)^2$:

$$V(x) = \int_{x_{min}}^{x_{max}} (x - \mu_x)^2 p(x)dx. \tag{5.3}$$

Equation 5.3 can be expressed as

$$V(x) = \langle x^2 \rangle - \langle x \rangle^2. \tag{5.4}$$

It can be shown that $V(A) = 0$, $V(Ax) = A^2 V(x)$, and $V(x + A) = V(x)$ if A is a constant. The standard deviation σ_x is simply the square root of the variance:

$$\sigma_x = \sqrt{V(x)}. \tag{5.5}$$

In many situations, we are interested in multiple random variables; e.g., a measured variable might be influenced by several error sources, which are modeled as random variables. For simplicity, consider first only two random variables x and y, which have the joint pdf $p(x, y)$. The expectation of $g(x, y)$ is given by

$$\langle g(x,y) \rangle = \int_{y_{min}}^{y_{max}} \int_{x_{min}}^{x_{max}} g(x,y)p(x,y)dxdy. \tag{5.6}$$

In particular, the mean values for x and y are

$$\mu_x = \int_{y_{min}}^{y_{max}} \int_{x_{min}}^{x_{max}} xp(x,y)dxdy \tag{5.7}$$

and

$$\mu_y = \int_{y_{min}}^{y_{max}} \int_{x_{min}}^{x_{max}} yp(x,y)dxdy \tag{5.8}$$

and their corresponding variances are

$$V(x) = \int\limits_{y_{min}}^{y_{max}} \int\limits_{x_{min}}^{x_{max}} (x - \mu_x)^2 \, p(x,y) \, dxdy \tag{5.9}$$

and

$$V(y) = \int\limits_{y_{min}}^{y_{max}} \int\limits_{x_{min}}^{x_{max}} (y - \mu_y)^2 \, p(x,y) \, dxdy. \tag{5.10}$$

If a random variable z is defined as the sum

$$z = Ax + By, \tag{5.11}$$

where
  A and B are constants
  x and y are random variables, it can be shown that the following expressions apply:

$$\mu_z = \langle z \rangle = \langle Ax + By \rangle = A\mu_x + B\mu_y \tag{5.12}$$

and

$$V(z) = \langle (z - \mu_z)^2 \rangle = A^2 V(x) + B^2 V(y) + 2ABC(x,y). \tag{5.13}$$

The term $C(x, y)$ is the covariance between x and y and is defined as

$$C(x,y) = \langle (x - \mu_x)(y - \mu_y) \rangle. \tag{5.14}$$

Note that $C(x, x) = V(x)$; i.e., the variance of a random variable is the covariance of a random variable with itself. Equation 5.13 can thus be written as

$$V(z) = A^2 C(x,x) + B^2 C(y,y) + 2ABC(x,y). \tag{5.15}$$

Equation 5.15 can be extended to the case where z is a linear combination of any number of random variables x_i:

$$z = A_1 x_1 + A_2 x_2 + \cdots + A_J x_J = \sum_{i=1}^{J} A_i x_i, \tag{5.16}$$

leading to

$$V(z) = \sum_{i=1}^{J} \sum_{j=1}^{J} A_i A_j C(x_i, x_j).$$ (5.17)

Equation 5.17 is a key result for uncertainty analysis because it allows for an estimate of the dispersion of the variable z; i.e., the standard deviation of z is the square root of Equation 5.17.

The correlation coefficient $\rho_{x_i x_j}$ is defined in

$$\rho_{x_i x_j} = \frac{C(x_i, x_j)}{\left(V(x_i) V(x_j)\right)^{1/2}},$$ (5.18)

allowing Equation 5.17 to be expressed as

$$V(z) = \sum_{i=1}^{J} \sum_{j=1}^{J} A_i A_j \rho_{x_i x_j} \left(V(x_i) V(x_j)\right)^{1/2}.$$ (5.19)

It can be shown that $\rho_{x_i x_j}$ is always within the range -1 to 1 and that $\rho_{x_i x_i} = 1$. If $\rho_{x_i x_j} = 0$, the variables x_i and x_j are said to be uncorrelated.

Finally, we note that the variance of the product of two independent (uncorrelated) random variables x and y is given by [6]

$$V(xy) = \langle x \rangle^2 V(y) + \langle y \rangle^2 V(x) + V(x) V(y).$$ (5.20)

Equation 5.20 can be extended to account for correlations [6].

5.3 Measurement Uncertainties

Now consider the errors that can arise during measurement of a quantity x. If we denote random errors with the symbol ε and systematic errors with the symbol β, we can write

$$x = x_{\text{true}} + \varepsilon_1 + \varepsilon_2 + \cdots + \varepsilon_L + \beta_1 + \beta_2 + \cdots + \beta_M.$$ (5.21)

Here, x_{true} is the true value we want to know (but cannot) and the subscripts on ε and β denote different identifiable errors (also called elemental errors) that we treat as random variables. We will assume that x_{true} is a random variable such that $x_{\text{true}} = \langle x \rangle + \alpha$, where α accounts for variations in the measurand itself. Let $\varepsilon = \varepsilon_1 + \varepsilon_2 + \cdots + \varepsilon_L$ and $\beta = \beta_1 + \beta_2 + \cdots + \beta_M$ so that Equation 5.21 can be written as Equation 5.22:

$$x = \langle x \rangle + \alpha + \varepsilon + \beta. \tag{5.22}$$

The variance of Equation 5.22 is

$$V(x) = V(\alpha) + V(\varepsilon) + V(\beta) + 2C(\varepsilon,\beta) + 2C(\varepsilon,\alpha) + 2C(\alpha,\beta). \tag{5.23}$$

Note that $\langle x \rangle$ is assumed to be constant. The covariance terms in Equation 5.23 actually contain multiple terms and can be written as sums of covariances, e.g.,

$$C(\varepsilon,\beta) = C(\varepsilon_1 + \varepsilon_2 + \cdots + \varepsilon_L, \beta_1 + \beta_2 + \cdots + \beta_M) = \sum_{i=1}^{L}\sum_{j=1}^{M} C(\varepsilon_i,\beta_j).$$

If we assume that the random errors, the systematic errors, and α are all uncorrelated and define the variables

$$u_x = \sqrt{V(x)}, \tag{5.24}$$

$$s_x = \sqrt{V(\alpha) + V(\varepsilon)}, \tag{5.25}$$

$$b_x = \sqrt{V(\beta)}, \tag{5.26}$$

then we get

$$u_x = \sqrt{s_x^2 + b_x^2}. \tag{5.27}$$

The term b_x accounts for systematic errors and u_x is termed the standard systematic uncertainty in x. The word "standard" is used here because u_x is the square root of a variance, which is a standard deviation. The term s_x accounts for random variations in the quantity being measured with $V(x_{\text{true}})$ as well as random errors associated with the measurement system with $V(\varepsilon)$.

The earlier expressions apply to populations. In practice, s_x is typically estimated by evaluating the standard deviation of a sample:

$$s_x = \left(\frac{1}{N-1} \sum_{i=1}^{N} (x_i - \bar{x})^2 \right)^{1/2}. \tag{5.28}$$

The s_x term can also be estimated by applying Equation 5.17 to the elemental random error terms (if they can be identified), as shown in

$$s_x = \left(V(\alpha) + \sum_{i=1}^{L} \sum_{j=1}^{L} C(\varepsilon_i, \varepsilon_j) \right)^{1/2}, \tag{5.29}$$

but it can be difficult to properly account for covariances between random errors. Equation 5.28 automatically includes estimates of the effects of covariances. If all random errors are independent of each other (uncorrelated), then Equation 5.29 becomes

$$s_x = \sqrt{V(\alpha) + s_1^2 + s_2^2 + \cdots + s_L^2}, \tag{5.30}$$

where
$s_i^2 = V(\varepsilon_i)$ is the variance of elemental random error i
s_i is the corresponding standard deviation

The b_x term is typically estimated by assuming that the systematic errors are uncorrelated, which yields

$$b_x = \sqrt{b_1^2 + b_2^2 + \cdots + b_M^2}, \tag{5.31}$$

where $b_i^2 = V(\beta_i)$ is the variance of elemental systematic error i. It thus follows that b_i is the standard deviation of elemental systematic error i. Often, β_i is a constant or is assumed to be proportional to the value being read, i.e., $\beta_i = \phi_i x$, where ϕ_i is a random variable. For this situation,

$$V(\beta_i) = \langle x \rangle^2 V(\phi_i) + \langle \phi_i \rangle^2 V(x) + V(\phi_i) V(x)$$

from Equation 5.20. Because $\langle x \rangle$ is not known, it can be estimated as $\bar{x}$. It can often be reasonably assumed that $\langle \phi_i \rangle = 0$ as well.

Equation 5.30 states that b_x is the square root of the sum of variances of the elemental systematic errors. Neglecting systematic error correlations works well for estimating uncertainties of measured values, but it can lead to serious errors when measured values are used to calculate other results, as discussed in a later chapter. Note that, to use Equations 5.29 through 5.31, we do not need to know the underlying distributions for the elemental errors, and in fact these distributions can all be different or the same.

Equation 5.27 provides an estimate of the spread in the data that we would obtain if we changed the instrumentation every time a new measurement was obtained, but where the exact same model, type, manufacturer, etc., for the instrumentation is employed to obtain each data point. This is what allows us to treat the systematic errors as random errors. We are assuming that there is a population of instruments that has a population of systematic errors that vary randomly among the instruments.

In many cases we want to know the uncertainty of a mean value obtained from a series of N measurements with the same instrumentation. This can be estimated by using the procedures described earlier, but where our random variable is the data mean,

$$\bar{x} = \frac{x_1 + x_2 + \cdots + x_N}{N} = \frac{x_1}{N} + \frac{x_2}{N} + \cdots + \frac{x_N}{N} = \sum_{i=1}^{N} \frac{x_i}{N}. \tag{5.32}$$

Each measurement x_i will in general have a different random error ε_i and a different fluctuation α_i in the measurand, but the systematic error β will be unchanged, i.e.,

$$x_i = \langle x \rangle + \alpha_i + \varepsilon_i + \beta. \tag{5.33}$$

Substituting Equation 5.33 into Equation 5.32 yields

$$\bar{x} = \langle x \rangle + \sum_{i=1}^{N} \frac{\alpha_i}{N} + \sum_{i=1}^{N} \frac{\varepsilon_i}{N} + \beta. \tag{5.34}$$

The summation involving ε_i and α_i in Equation 5.34 will become small as N becomes large such that these terms will become negligible. However, the β term is independent of N; i.e., averaging cannot reduce systematic errors. Of course, this is true only if the instrumentation is unchanged.

We now take the variance of Equation 5.34, which yields

$$V(\bar{x}) = \frac{V(\alpha)}{N} + \frac{V(\varepsilon)}{N} + V(\beta). \tag{5.35}$$

If we define $u_{\bar{x}} = \sqrt{V(\bar{x})}$ as the standard uncertainty of the mean and note that the standard error of the mean is defined as $s_{\bar{x}} = s_x / \sqrt{N}$, we obtain

$$u_{\bar{x}} = \sqrt{s_{\bar{x}}^2 + b_{\bar{x}}^2}. \tag{5.36}$$

The variable $b_{\bar{x}}$ is the standard deviation associated with systematic errors. It turns out that $b_{\bar{x}} = b_x$.

The expanded uncertainty is defined to be the standard uncertainty multiplied by a coverage factor k. For example, for uncertainties in $\bar{x}$, the following equation would apply:

$$U_{\bar{x}} = k u_{\bar{x}} = k \sqrt{s_{\bar{x}}^2 + b_{\bar{x}}^2} \quad (P\%). \tag{5.37}$$

The value of k depends on factors such as the confidence level P and the number of data points. We will discuss the coverage factor in detail later in the chapter, but it typically turns out that $k = 2$ can be assumed to correspond to $P = 0.95$.

The relative uncertainty of the mean is simply a variable's expanded uncertainty of the mean divided by its mean value:

$$\text{Relative uncertainty} = \frac{U_{\bar{x}}}{\bar{x}} \quad (P\%).$$

Knowing the uncertainty of a measurement will tell us something about an *interval* within which the true value of the measured variable is likely to be found. We are interested in evaluating uncertainty intervals of the form

$$\text{Uncertainty interval} = \text{middle value} \pm \text{uncertainty} \quad (P\%).$$

Note that we have used the term "uncertainty interval" here instead of "confidence interval," as the latter term applies when only random errors are present [3]. In particular, we are interested in evaluating the uncertainty interval for the mean of a population. The uncertainty interval for the true mean value is given by

$$\mu = \bar{x} \pm U_{\bar{x}} \quad (P\%). \tag{5.38}$$

5.4 Elemental Systematic Errors

Systematic errors are assumed to always occur in the same way during a measurement, so it can be difficult to know whether or not they are even present. One approach to addressing this issue, however, is to assume that systematic errors *can* be present with certain probabilities, which allows us to use statistical methods to evaluate them.

An elemental systematic error is an identifiable individual error that contributes to the overall systematic error. Elemental errors can often be characterized in some sense by considering factors such as information from the equipment manufacturer, experience with similar instruments, calibration uncertainties, etc. For example, a manufacturer might state that a voltage measurement system has a linearity of ±0.1 mV. What this means is that the manufacturer would (should) have investigated a large number of instruments (all of the same model) and assessed the linearity of each instrument. The ±0.1 mV value would result from the manufacturer's statistical analysis of such an investigation. This value is actually a confidence interval, and if we convert it into a standard deviation, we obtain the elemental error standard deviation "b" that we need for the uncertainty analysis.

To convert this voltage linearity value into a standard deviation, we need to know its probability distribution and also the applicable confidence level. There are a few typical probability distributions to consider: (1) normal, (2) uniform, and (3) triangular. We consider these different possibilities next for a linearity of ±0.1 mV.

5.4.1 Normal Distributions

If the manufacturer does not give any information about the distribution or the confidence level, it is recommended to assume that the distribution is normal and that the confidence level is 95%. In this case, you can calculate the b value simply by dividing the confidence interval by 1.96, which we just rounded to 2, yielding a standard deviation of $b = 0.05$ mV. A plot of this distribution can be made using the following R code:

```
> x <- seq(from=-0.2,to=+0.2,by=0.001)
> plot(x,dnorm(x,mean=0,sd=0.05),type="l")
```

The resulting plot is shown in Figure 5.1.

5.4.2 Uniform Distributions

Uniform distributions are used when it is stated that the errors can be anywhere between two values x_{min} and x_{max} with an equal probability and where

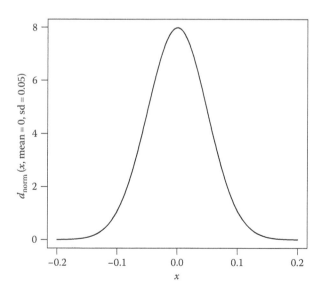

FIGURE 5.1
Normal distribution with a mean of 0 and a standard deviation of 0.05.

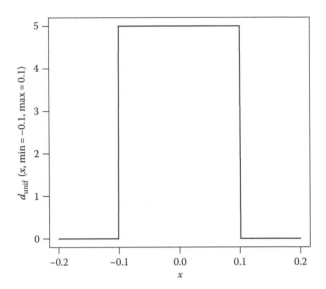

FIGURE 5.2
Uniform distribution with a mean of 0 and limits of −0.1 and 0.1.

no particular data range is favored. In this case, the standard deviation is $b = (x_{max} - x_{min})/\sqrt{12}$. For the voltage measurement system described earlier, this yields $b = 0.058$ mV, where $x_{min} = -0.1$ mV and $x_{max} = 0.1$ mV. A plot of this distribution obtained with the R code

```
> x <- seq(from=-0.2,to=+0.2,by=0.001)
> plot(x,dunif(x,min=-0.1,max=0.1),type="l")
```

is shown in Figure 5.2.

5.4.3 Triangular Distributions

Triangular distributions are used when it is stated that the errors can be anywhere between two values x_{min} and x_{max} but where it is more likely that errors will be close to the middle of this interval than further away. The standard deviation for this case is $b = (x_{max} - x_{min})/\sqrt{24}$. For the voltage measurement system described earlier, this yields $b = 0.041$ mV, where $x_{min} = -0.1$ mV and $x_{max} = 0.1$ mV. We have assumed that the triangular distribution is symmetric. A plot of this distribution obtained with the R code

```
> library(triangle)
> x <- seq(from=-0.2,to=+0.2,by=0.001)
> plot(x,dtriangle(x,a=-0.1,b=0.1),type="l")
```

is shown in Figure 5.3.

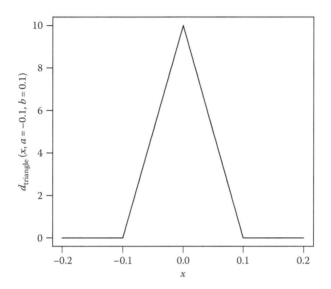

FIGURE 5.3
Symmetric triangular distribution with a mean of 0 and limits of −0.1 and 0.1.

5.5 Coverage Factors

The coverage factor k is clearly an important variable. One approach to calculating k is to assume

$$k = t_{v,1-\alpha/2}, \tag{5.39}$$

where v is the effective number of degrees of freedom, $\alpha = 1 - P$, and t is Student's t variable. The confidence level P is something we select. The effective number of degrees of freedom can be estimated using the Welch–Satterthwaite equation:

$$v = \frac{\left(s_x^2 + \sum_g b_{x,g}^2 \right)^2}{\dfrac{s_x^4}{v_{s_x}} + \sum_g \dfrac{b_{x,g}^4}{v_{b_{x,g}}}}. \tag{5.40}$$

The number of degrees of freedom associated with the variance of the measured data is $v_{s_x} = N - 1$. The number of degrees of freedom associated with the variance systematic error g, i.e., $b_{x,g}^2$, is denoted as $v_{b_{x,g}}$. This quantity can be difficult to evaluate as it is often only acquired with experience in using the instrumentation. A recommended approach, however, is to assume that

$$v_{b_{x,g}} = \frac{1}{2u_{x,g}^2},$$ (5.41)

where $u_{x,g}$ is the *relative uncertainty* of $b_{x,g}$. In the absence of other information, we will assume that these relative uncertainties have some reasonable value, e.g., 10%.

Consider the following example, where we have 10 data points, three systematic error sources, and the relative uncertainty of each error source is 10%:

```
> x <- c(23.9,25.3,24.8,26.8,25.2,25.6,23.8,26.4,23.3,24.3) # data
> b <- c(0.1,0.5,1)            # elemental systematic errors
> rel_b <- 0.1*c(1,1,1)        # relative systematic error uncertainties
> N <- length(x)               # number of data points
> # evaluate the Welch-Satterthwaite equation for the effective dof
> nu_top <- (sd(x)^2+sum(b^2))^2   # numerator
> nu_bot <- (sd(x)^4)/(N-1)+sum((b^4)/(0.5/(rel_b^2)))   # denominator
> nu <- floor(nu_top/nu_bot)   # calculate the effective dof
> nu                           # print the effective dof
[8] 31
```

The effective number of degrees of freedom is then $v = 31$ (rounded down to the nearest integer). For a 95% confidence level, we then calculate the coverage factor:

```
> qt(p=0.975,df=31)
[8] 2.039513
```

Therefore, $k = 2.04$.

In most situations, we will not know the relative uncertainties of the elemental systematic errors. However, it is often the case that $v \gg 1$ such that k can be considered to be independent of v. To see this, consider the plot of the $P = 95\%$ coverage factor $k (= t_{v,1-\alpha/2})$ in Figure 5.4. The solid line is the coverage factor and the dashed line is the asymptotic limit in k for $v \to \infty$. Here is the code used to make the plot:

```
> x <- seq(from=2,to=30,by=0.1)
> P <- 0.975
> plot(x,qt(p=P,df=x),type="l",col="black",ylim=c(0,5),xlab="Degrees of
  Freedom",ylab="Coverage Factor",main="P = 0.95")
> abline(h=1.96,lty=2)
```

For v greater than about 10, k is close to 2, and for $v \to \infty$, k approaches a value of 1.96. As a result, it is reasonable to just use $k = 2$ for $P = 95\%$ as long

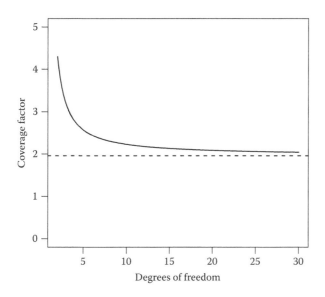

FIGURE 5.4
The 95% coverage factor as a function of the number of degrees of freedom.

as $\nu > 10$ (roughly). This criterion is typically satisfied in experiments, and unless otherwise stated, we will assume it applies.

As an example, we will use the following R code to evaluate the standard uncertainty (for the mean) for a set of measured temperatures:

```
> x <- c(23.9,25.3,24.8,26.8,25.2,25.6,23.8,26.4,23.3,24.3) # data
> b <- c(0.1,1,0.5)            # elemental systematic errors
> N <- length(x)               # number of data points
> stdMeanError <- sd(x)/sqrt(N)    # standard error of the mean
> u_xbar <- sqrt(stdMeanError^2 + sum(b^2))  # std uncertainty
> xbar <- mean(x)              # data mean
> k <- 1.96                    # coverage factor
> U_xbar <- k*u_xbar           # expanded uncertainty
> xmin <- xbar-U_xbar          # confidence interval lower limit
> xmax <- xbar+U_xbar          # confidence interval upper limit
> xmin                         # print the conf interval lower limit
[8] 22.62897
> xbar                         # print the data mean
[8] 24.94
> xmax                         # print the confidence interval upper
  limit
[8] 27.25103
> u_xbar                       # print the standard uncertainty
[8] 1.179096
> U_xbar                       # print the expanded uncertainty
[8] 2.311028
```

We will also evaluate an uncertainty interval for $k = 2$, which we assume here corresponds to $P = 0.95$. We have three elemental systematic error sources with standard deviations $b_{\bar{x},1} = 0.1$ K, $b_{\bar{x},2} = 1$ K, and $b_{\bar{x},3} = 0.5$ K. The uncertainty interval for the mean temperature is thus

$$\mu = 24.9 \pm 2.4 \text{ K} \ (95\%).$$

We plot the data in Figure 5.5 and also show the data mean (as a solid line). The dashed lines in Figure 5.5 denote the 95% uncertainty interval for the true mean of the data. Here is the code:

```
> plot(x,ylim=c(20,30))
> abline(h=mean(xmin),col="black")
> abline(h=mean(x),col="black",Hy="dashed")
> abline(h=mean(xmax),col="black",lty="dashed")
```

Calculations show that $s_{\bar{x}}^2 = 0.130$ and $b_{\bar{x}}^2 = 1.26$ such that the uncertainty is dominated by the systematic errors. If we wanted to substantially reduce the uncertainty, we would need to obtain better instrumentation. Simply obtaining more data with the current instrumentation would not help very much in this regard.

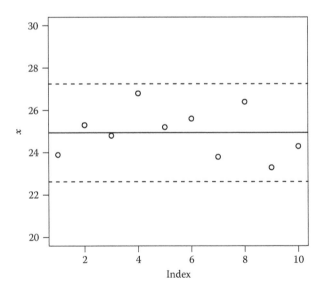

FIGURE 5.5
Plot of the data (circles), the data mean (solid line), and the 95% uncertainty interval for the data mean (dashed lines).

Problems

P5.1 Evaluate the correlation coefficient for the situation where $p(x, y) = g(x)h(y)$, i.e., $p(x, y)$ is the product of the pdfs $g(x)$ and $h(y)$.

P5.2 A normal distribution, a uniform distribution, and a symmetric triangular distribution each have a standard deviation of 1. What is the probability that a random value will be within plus or minus one standard deviation of the mean for each distribution?

P5.3 The independent random variables W, X, and Y have a normal distribution, a uniform distribution, and a symmetric triangular distribution, respectively, with $\mu_X = 1$, $\mu_Y = -1$, $\mu_W = 2$, $\sigma_X = 1$, $\sigma_Y = 2$, and $\sigma_W = 3$. Calculate the mean and standard deviation of the random variable Z, where $Z = W + 2X - 2Y$. Do this by using the analytical expressions in this chapter as well as numerically using R code. In the R code, investigate different values of the number of random values, N, that are used when numerical values are calculated.

P5.4 Show the steps required to go from Equations 5.29 to 5.30.

P5.5 Show the steps required to go from Equations 5.34 to 5.35.

P5.6 The temperature of a liquid is considered to be a random variable with a uniform pdf. The minimum temperature is 303 K and the maximum is 307 K. Using R code, simulate the data that would be obtained if N measurements of the temperature were obtained for $N = 10$, 10^3, and 10^5. In particular, evaluate the mean and standard deviation for each value of N. Repeat the simulations M times, where $M = 10^5$, and plot a histogram of the values for the mean and the standard deviation. Calculate a standard deviation for the mean values and compare it with an analytical prediction of the standard deviation of the mean. (By doing all of this, you have numerically checked the applicability of the central limit theorem.)

P5.7 Derive Equation 5.17.

P5.8 A temperature measurement system has three elemental systematic errors with $b_1 = 0.5$ K, $b_2 = 1.5$ K, and $b_3 = 2.2$ K. Random variations are negligible, i.e., $s_x = 0$. Calculate the coverage factor, assuming that the relative uncertainties in the systematic errors are all 10%. Repeat your calculations using relative uncertainties of 1% and 50%. Comment on any trends in your results.

P5.9 A pressure measurement system has two systematic errors: $\beta_1 = \phi_1 p$ and $\beta_2 = $ constant, where p is the absolute pressure in kilopascals. The standard deviation for ϕ_1 is 0.01 and the standard deviation for β_2 is 0.2 kPa. A set of $N = 20$ measurements yields a mean pressure of

20.1 kPa and a standard deviation of 0.3 kPa. Calculate the 95% uncertainty interval for the mean pressure.

P5.10 A temperature measurement system was used to measure the temperature of gases exiting a combustor. The measurement system has the following systematic errors: linearity = 2 K, hysteresis = 2 K, repeatability = 2 K, and resolution = 1 K. The results from 10 measurements are shown in the following table. Check the data for outliers using the modified Thompson tau test. After rejecting any outliers, calculate the 95% uncertainty interval for the mean gas temperature.

i	T_i (K)
1	1277
2	1264
3	1289
4	1223
5	1261
6	1231
7	1238
8	1258
9	1221
10	1230

References

1. ASME PTC 19.1-2013, Test uncertainty—Performance test codes, The American Society of Mechanical Engineers, New York, 2013.
2. H. W. Coleman and W. G. Steele, *Experimentation, Validation, and Uncertainty Analysis for Engineers*, 3rd edn., Wiley, Hoboken, NJ, 2009.
3. JCGM 100:2008, Evaluation of measurement data—Guide to the expression of uncertainty in measurement, GUM 1995 with minor corrections, International Bureau of Weight and Measures (BIPM), Sérres, France, 2008.
4. JCGM 101:2008, Evaluation of measurement data: Supplement 1 to the "Guide to the expression of uncertainty in measurement"—Propagation of distributions using a Monte Carlo method, International Bureau of Weight and Measures (BIPM), Sérres, France, 2008.
5. B. N. Taylor and C. E. Kuyatt, Guidelines for evaluating and expressing the uncertainty of NIST measurement results, NIST Technical Note 1297, 1994 Edition (Supersedes 1993 Edition), National Institute of Standards and Technology, Gaithersburg, MD, 1994.
6. L. A. Goodman, On the exact variance of products, *Journal of the American Statistical Association* 55, 708–713, 1960.

6

Uncertainty of a Result Calculated Using Experimental Data

Oftentimes, we use results from experiments to calculate other quantities [1–4]. For example, if we measure the temperature and pressure of a gas, we could calculate its density using an equation of state. A basic question then is, "What is the uncertainty of a calculated result if we account for the uncertainties of the input quantities?" To address this question we will use *uncertainty propagation*; i.e., we propagate experimental uncertainties into a calculated result.

Suppose we want to calculate a result r that is a function of L random variables $x_1, x_2, \ldots, x_L$:

$$r = r(x_1, x_2, \ldots, x_L). \tag{6.1}$$

The random variables are assumed to have probability density functions that are independent of time.

The standard uncertainty in r, denoted as u_r, is the square root of the variance of Equation 6.1, that is,

$$u_r = \sqrt{V(r)}. \tag{6.2}$$

The variance of r is given by

$$V(r) = \langle r^2 \rangle - \langle r \rangle^2, \tag{6.3}$$

where the mean values $\langle r \rangle$ and $\langle r^2 \rangle$ are defined by

$$\langle r \rangle = \int_{x_{min,L}}^{x_{max,L}} \cdots \int_{x_{min,1}}^{x_{max,1}} r(x_1, x_2, \ldots, x_L) p(x_1, x_2, \ldots, x_L) dx_1 dx_2 dx_L \tag{6.4}$$

and

$$\langle r^2 \rangle = \int_{x_{min,L}}^{x_{max,L}} \cdots \int_{x_{min,1}}^{x_{max,1}} r^2(x_1, x_2, \ldots, x_L) p(x_1, x_2, \ldots, x_L) dx_1 dx_2 dx_L. \tag{6.5}$$

The integrals in Equations 6.4 and 6.5 can be difficult or impossible to evaluate exactly, so there are approximate approaches that are commonly employed. One of these approaches involves linearizing Equation 6.1 using a Taylor series. This is the approach we will use in this chapter. Another approximate approach, which is termed the Monte Carlo approach, is purely computational and will be discussed in a later chapter. Although the Monte Carlo approach is generally more robust than the Taylor series approach, it does not yield analytical expressions that can provide physical insight.

6.1 Taylor Series Approach

We will express Equation 6.1 as

$$r = r\left(\langle x_1 \rangle + \delta x_1, \ \langle x_2 \rangle + \delta x_2, \dots, \langle x_L \rangle + \delta x_L\right), \tag{6.6}$$

where $\delta x_m = x_m - \langle x_m \rangle$ and $x_m = \langle x_m \rangle + \alpha_m + \varepsilon_m + \beta_m$. The subscript m is an integer that takes on a value in the range 1, 2,..., L. The α_m term accounts for variations in the measurand x_m itself, while ε_m and β_m are the overall random and systematic errors that apply to measurements of x_m. Expanding Equation 6.1 in a Taylor series about the $\langle x_m \rangle$ values and neglecting terms of order δx_m^2 or higher yields

$$r = \langle r \rangle + \theta_1\left(\alpha_1 + \varepsilon_1 + \beta_1\right) + \theta_2\left(\alpha_2 + \varepsilon_2 + \beta_2\right) + \cdots + \theta_L\left(\alpha_L + \varepsilon_L + \beta_L\right), \tag{6.7}$$

where the mean value $\langle r \rangle$ is given by

$$\langle r \rangle = r\left(\langle x_1 \rangle, \langle x_2 \rangle, \dots, \langle x_L \rangle\right) \tag{6.8}$$

and the θ_m terms, commonly called sensitivity coefficients, are the partial derivatives defined in

$$\theta_m = \left. \frac{\partial r}{\partial x_m} \right|_{x_1 = \langle x_1 \rangle, \dots, \ x_L = \langle x_L \rangle}. \tag{6.9}$$

If we define the random variables α, ε, and β as

$$\alpha = \theta_1\alpha_1 + \theta_2\alpha_2 + \cdots + \theta_L\alpha_L, \tag{6.10}$$

$$\varepsilon = \theta_1\varepsilon_1 + \theta_2\varepsilon_2 + \cdots + \theta_L\varepsilon_L, \tag{6.11}$$

and

$$\beta = \theta_1 \beta_1 + \theta_2 \beta_2 + \cdots + \theta_L \beta_L, \tag{6.12}$$

we can express Equation 6.7 as

$$r = \langle r \rangle + \alpha + \varepsilon + \beta. \tag{6.13}$$

Equation 6.13 is advantageous because it is a linear sum of random variables with $\langle r \rangle$ considered as constant. The variance of r can thus be estimated by using

$$V(r) = V(\alpha) + V(\varepsilon) + V(\beta) + 2C(\alpha, \varepsilon) + 2C(\alpha, \beta) + 2C(\varepsilon, \beta). \tag{6.14}$$

To continue, we assume that α, ε, and β are all independent of each other. The covariances between these variables in Equation 6.14 are then zero, yielding

$$V(r) = V(\alpha) + V(\varepsilon) + V(\beta) \tag{6.15}$$

for the variance of r. We now define the quantities s_r^2, b_r^2, and u_r^2 as

$$s_r^2 = V(\alpha) + V(\varepsilon), \tag{6.16}$$

$$b_r^2 = V(\beta), \tag{6.17}$$

and

$$u_r^2 = s_r^2 + b_r^2. \tag{6.18}$$

The quantity u_r^2 is the variance in r such that the standard uncertainty in r is then given by

$$u_r = \sqrt{s_r^2 + b_r^2}. \tag{6.19}$$

To evaluate the standard uncertainty of the mean of r, i.e., $u_{\bar{r}}$, consider the mean value $\bar{r}$ defined in

$$\bar{r} = \frac{1}{N} \sum_{i=1}^{N} r_i. \tag{6.20}$$

The r_i terms in Equation 6.20 refer to individual values of r that would be calculated using N measured values of each of the random variables $x_1, x_2, \ldots, x_L$, that is,

$$r_i = \langle r \rangle + \alpha_i + \varepsilon_i + \beta. \tag{6.21}$$

Inserting Equation 6.21 into Equation 6.20 yields

$$\bar{r} = \langle r \rangle + \sum_{i=1}^{N} \frac{\alpha_i}{N} + \sum_{i=1}^{N} \frac{\varepsilon_i}{N} + \beta. \tag{6.22}$$

The variance of Equation 6.22 is given by

$$V(\bar{r}) = \frac{V(\alpha) + V(\varepsilon)}{N} + V(\beta), \tag{6.23}$$

which can be written as

$$V(\bar{r}) = \frac{s_r^2}{N} + V(\beta). \tag{6.24}$$

The ratio $s_r / \sqrt{N}$ is the standard deviation of the mean of r, and we denote this as $s_{\bar{r}}$. We also note that $b_{\bar{r}} = b_r = \sqrt{V(\beta)}$ such that the standard uncertainty $u_{\bar{r}}$ is given by

$$u_{\bar{r}} = \sqrt{V(\bar{r})} = \sqrt{s_{\bar{r}}^2 + b_{\bar{r}}^2}. \tag{6.25}$$

Equations 6.19 and 6.25 apply to populations, so we need to discuss how they are applied to samples. The mean values $\langle x_m \rangle$ are estimated as their sample means $\bar{x}_m$ and the sensitivity coefficients are evaluated by using the $\bar{x}_m$ terms. The variances in Equation 6.15 are the following sums of covariances:

$$V(\alpha) = \sum_{m=1}^{L} \sum_{m=1}^{L} \theta_m \theta_n C(\alpha_m, \alpha_n), \tag{6.26}$$

$$V(\varepsilon) = \sum_{m=1}^{L} \sum_{n=1}^{L} \theta_m \theta_n C(\varepsilon_m, \varepsilon_n), \tag{6.27}$$

and

$$V(\beta) = \sum_{m=1}^{L} \sum_{n=1}^{L} \theta_m \theta_n C(\beta_m, \beta_n). \tag{6.28}$$

We define the terms s_{mn} and b_{mn} as

$$s_{mn} = \frac{C(\alpha_m, \alpha_n) + C(\varepsilon_m, \varepsilon_n)}{N} \tag{6.29}$$

and

$$b_{mn} = C(\beta_m, \beta_n) \tag{6.30}$$

such that

$$s_r^2 = \sum_{m=1}^{L}\sum_{n=1}^{L}\theta_m\theta_n s_{mn} \tag{6.31}$$

and

$$b_r^2 = \sum_{m=1}^{L}\sum_{m=1}^{L}\theta_m\theta_n b_{mn}. \tag{6.32}$$

Equations 6.31 and 6.32, when written out for a few values of L, are as follows:

$L = 1: s_r^2 = \theta_1^2 s_{11},$
$L = 2: s_r^2 = \theta_1^2 s_{11} + \theta_2^2 s_{22} + 2\theta_1\theta_2 s_{12},$
$L = 3: s_r^2 = \theta_1^2 s_{11} + \theta_2^2 s_{22} + \theta_3^2 s_{33} + 2\theta_1\theta_2 s_{12} + 2\theta_1\theta_3 s_{13} + 2\theta_2\theta_3 s_{23},$
$L = 1: b_r^2 = \theta_1^2 b_{11},$
$L = 2: b_r^2 = \theta_1^2 b_{11} + \theta_2^2 b_{22} + 2\theta_1\theta_2 b_{12},$
$L = 3: b_r^2 = \theta_1^2 b_{11} + \theta_2^2 b_{22} + \theta_3^2 b_{33} + 2\theta_1\theta_2 b_{12} + 2\theta_1\theta_3 b_{13} + 2\theta_2\theta_3 b_{23}.$

The variables s_{mn} and b_{mn} are to be treated as follows:

1. s_{mm} is the square of the random standard deviation of the mean of variable m, i.e., $s_{mm} = s_{x_m}^2 / N = s_{\bar{x}_m}^2$.
2. s_{mn} is the covariance of $\bar{x}_m$ and $\bar{x}_n$.
3. b_{mm} is the sum of the variances of the elemental systematic errors of $\bar{x}_m$.
4. b_{mn} is the sum of the covariances of the elemental systematic errors that are *common* to $\bar{x}_m$ and $\bar{x}_n$.
5. $b_{mn} = b_{nm}$ and $s_{mn} = s_{nm}$.

The covariance of $\bar{x}_m$ and $\bar{x}_n$ is calculated using

$$s_{mn} = \frac{1}{N(N-1)}\sum_{i=1}^{N}(x_{m,i} - \bar{x}_m)(x_{n,i} - \bar{x}_n). \tag{6.33}$$

If $s_{mn} = 0$, then there is no correlation between $\bar{x}_m$ and $\bar{x}_n$. Even if two variables are not correlated, random variations can produce data once in a while that result in a calculated covariance (s_{mn}) that is not small. This can especially be the case for small samples. As a result, if you are sure that $\bar{x}_m$ and

$\bar{x}_n$ cannot be correlated, you can just set $s_{mn} = 0$. An example could be where temperature is measured using a measurement system and pressure is measured using an entirely different measurement system that is completely uncoupled from the temperature measurements. If you do set $s_{mn} = 0$, however, it is advisable to also include results using Equation 6.33 and report your findings for both the cases.

The term $s_{\bar{r}}^2$ can also be calculated using the "direct approach" [4]:

$$s_{\bar{r}}^2 = \frac{1}{N-1}\sum_{i=1}^{N}\left(r_i - \bar{r}\right)^2, \tag{6.34}$$

where the summation is over the N data-tuples. Once we have calculated $s_{\bar{r}}^2$ in this way, we can calculate the standard deviation of the mean of r as $s_{\bar{r}} = s_r/\sqrt{N}$. This approach automatically includes covariance effects and may be preferable if it is difficult to characterize any of the covariance terms in Equation 6.29.

As an example, consider a situation where we measure the lengths of two sides, x and y. Side x is measured 10 times using a set of calipers and side y is measured 10 times using a *different* set of calipers. The resulting data are listed in Table 6.1. We want to use this information to calculate the mean of the length difference $r = y - x$ as well as its 95% confidence interval.

Our measurement systems are known to have the following elemental systematic errors:

Caliper *x*	
Linearity	0.01 mm (95%)
Hysteresis	0.005 mm (95%)
Calibration	0.005 mm (95%)

Caliper *y*	
Linearity	0.01 mm (95%)
Hysteresis	0.005 mm (95%)
Calibration	0.005 mm (95%)

For this example, both calipers were calibrated using the *same* standard, so this elemental error is *common* to both sets of measurements and

TABLE 6.1

Example Data of Length Measurements

x (mm)	9.93	9.91	10.14	10.19	10.00	10.01	10.02	10.03	10.11	10.01
y (mm)	19.87	20.07	20.10	19.98	20.21	20.07	19.88	20.08	20.01	19.86

will contribute to b_{12}. The other elemental errors are not common and will not contribute to b_{12}. We will solve this problem by hand as well as with R code.

For this problem, $L = 2$ and the expressions for s_r^2 and b_r^2 are

$$s_r^2 = \theta_1^2 s_{11} + \theta_2^2 s_{22} + 2\theta_1 \theta_2 s_{12}$$

and

$$b_r^2 = \theta_1^2 b_{11} + \theta_2^2 b_{22} + 2\theta_1 \theta_2 b_{12},$$

where $x_1 = x$ and $x_2 = y$, $\theta_1 = \theta_x = -1$, and $\theta_2 = \theta_y = +1$. We can also calculate

$$\bar{x}_1 = 10.035 \text{ mm},$$

$$\bar{x}_2 = 20.013 \text{ mm},$$

$$\bar{r} = 9.978 \text{ mm},$$

$$s_{11} = s_1^2 / N = 0.000778 \text{ mm}^2,$$

$$s_{22} = s_2^2 / N = 0.00133 \text{ mm}^2,$$

$$s_{12} = 0 \text{ (assumed)},$$

$$s_r^2 = \theta_1^2 s_{11} + \theta_2^2 s_{22} + 2\theta_1 \theta_2 s_{12} = 0.00211 \text{ mm}^2,$$

$$s_{\bar{r}} = \sqrt{s_r^2} = 0.0459 \text{ mm},$$

$$b_{11} = \left(\frac{0.01}{2}\right)^2 + \left(\frac{0.005}{2}\right)^2 + \left(\frac{0.005}{2}\right)^2 = 0.0000375 \text{ mm}^2,$$

$$b_{22} = \left(\frac{0.01}{2}\right)^2 + \left(\frac{0.005}{2}\right)^2 + \left(\frac{0.005}{2}\right)^2 = 0.0000375 \text{ mm}^2,$$

$$b_{12} = \left(\frac{0.005}{2}\right)^2 = 0.00000625 \text{ mm}^2,$$

$$b_r^2 = \theta_1^2 b_{11} + \theta_2^2 b_{22} + 2\theta_1 \theta_2 b_{12} = 0.0000625 \text{ mm}^2,$$

$$b_{\bar{r}} = \sqrt{b_r^2} = 0.0079 \text{ mm},$$

$$u_{\bar{r}} = \sqrt{s_r^2 + b_r^2} = 0.0466 \text{ mm},$$

and

$$U_{\bar{r}} = ku_{\bar{r}} = 0.0933 \text{ mm}.$$

The confidence interval for $\bar{r}$ is then

$$\bar{r} = 9.98 \pm 0.09 \text{ mm } (95\%).$$

We assumed that $s_{12} = 0$ because we did see a physical reason as to why random errors for the two calipers would be correlated.

The R code for this problem is as follows:

```
> x <- c(9.93,9.91,10.14,10.19,10.00,10.01,10.02,10.03,10.11,10.01)
> y <- c(19.87,20.07,20.10,19.98,20.21,20.07,19.88,20.08,20.01,19.86)
> xBar <- mean(x)
> yBar <- mean(y)
> theta1 <- -1
> theta2 <- +1
> rBar <- yBar-xBar
> bLin <- 0.01/2          # linearity
> bHys <- 0.005/2         # hysteresis
> bCal <- 0.005/2         # calibration
>
> b11 <- bLin^2 + bHys^2 + bCal^2
> b22 <- bLin^2 + bHys^2 + bCal^2
> b12 <- bCal^2
> brBar <- sqrt(theta1^2*b11+theta2^2*b22+2*theta1*theta2*b12)
>
> s11 <- var(x)/length(x)    # variance of xBar
> s22 <- var(y)/length(y)    # variance of yBar
> s12 <- 0                   # assumed covariance of xBar and yBar
> srBar <- sqrt(theta1^2*s11+theta2^2*s22+2*theta1*theta2*s12)
>
> urBar <- sqrt(srBar^2 + brBar^2)
> k <- 2
> UrBar <- k*urBar
> srBar
[1] 0.04595408
> brBar
[1] 0.007905694
> urBar
[1] 0.04662915
> rBar
[1] 9.978
> UrBar
[1] 0.0932583
```

So we obtain

$$\bar{r} = 9.98 \pm 0.09 \text{ mm } (95\%),$$

in agreement with the hand-calculated results.

We could have instead calculated $s_{\bar{r}}$ using the direct approach. This can be done with the following R code:

```
> x <- c(9.93,9.91,10.14,10.19,10.00,10.01,10.02,10.03,10.11,10.01)
> y <- c(19.87,20.07,20.10,19.98,20.21,20.07,19.88,20.08,20.01,19.86)
> rCalc <- y-x
> srBar <- sd(rCalc)/sqrt(length(rCalc))
> srBar
[1] 0.04376198
```

The result, $s_{\bar{r}} = 0.044$ mm, is close to the previous calculation in which $s_{12} = 0$ was used, suggesting that random covariance effects are irrelevant. This can be further checked because s_{12} can be calculated as the covariance of x and y divided by the number of data pairs (10 in this case). Here is the R code:

```
> x <- c(9.93,9.91,10.14,10.19,10.00,10.01,10.02,10.03,10.11,10.01)
> y <- c(19.87,20.07,20.10,19.98,20.21,20.07,19.88,20.08,20.01,19.86)
> cov(x,y)/10
[1] 9.833333e-05
```

This value for s_{12} is small compared with the calculated values $s_{11} = 0.0007783333$ and $s_{22} = 0.001333444$.

Finally, it should be mentioned that, if you have only *one* measurement of a random variable, you need to either neglect any terms related to random errors or estimate these terms using your own judgment or previous experimental data.

6.2 Coverage Factors

Similar to what we considered when the uncertainty of a measured variable was discussed, a typical approach for specifying the coverage factor, k, for a propagated uncertainty is to use

$$k = t_{v,(1+P)/2} \qquad (6.35)$$

where the effective number of degrees of freedom, v, can be estimated using the Welch–Satterthwaite equation.

Consider a variable Z as being the sum of variances:

$$Z = \sum_{i=1}^{L} c_i V_i, \qquad (6.36)$$

where
  V_i is a variance
  c_i is a constant

The effective number of degrees of freedom for Z according to the Welch–Satterthwaite equation is given by

$$\frac{Z^2}{\nu} = \sum_{i=1}^{L} \frac{(c_i V_i)^2}{\nu_i}, \tag{6.37}$$

where ν_i is the number of degrees of freedom associated with V_i. To use this expression in the present context, we define Z as

$$Z = \sum_{i=1}^{L} \theta_i^2 s_{ii} + \sum_{i=1}^{L} \theta_i^2 b_{ii}, \tag{6.38}$$

where we identify $c_i = \theta_i^2$ and $V_i = s_{ii}$ or b_{ii}. Recall that b_{ii} is the sum of the variances of the elemental systematic errors of $\bar{x}_i$. For clarity, we can define

$$b_{ii} = \sum_{k=1}^{K_i} b_{iik}, \tag{6.39}$$

where
  b_{iik} is the kth elemental variance for experimental variable i
  K_i is the total number of elemental systematic standard errors for variable i

Equation 6.38 then becomes

$$Z = \sum_{i=1}^{L} \theta_i^2 s_{ii} + \sum_{i=1}^{L} \theta_i^2 \left(\sum_{k=1}^{K_i} b_{iik} \right), \tag{6.40}$$

which is a sum of variances. We can thus write

$$\frac{Z^2}{\nu} = \sum_{i=1}^{L} \frac{\left(\theta_i^2 s_{ii} \right)^2}{\nu_{si}} + \sum_{i=1}^{L} \left(\sum_{k=1}^{K_i} \frac{\left(\theta_i^2 b_{iik} \right)^2}{\nu_{bik}} \right). \tag{6.41}$$

The variable ν_{si} is the number of degrees of freedom for s_{ii} and ν_{bik} is the number of degrees of freedom for b_{ik}. These quantities are generally calculated using

$$\nu_{si} = N - 1 \tag{6.42}$$

and

$$\nu_{bik} = \frac{1}{2} \left(\frac{\Delta b_{ik}}{b_{ik}} \right)^{-2}. \tag{6.43}$$

The ratio in parentheses in Equation 6.43 is an estimate of the relative uncertainty of b_{i_k}. A typical value might be $\Delta b_{i_k}/b_{i_k} \approx 0.25$, which corresponds to $\nu_{b_{i_k}} \approx 8$. As the relative uncertainty decreases, the number of degrees of freedom increases.

Consider the previous example where we used the measurements from two calipers to calculate a length difference. When we evaluated the expanded uncertainty, we used $k = 2$ for $P = 95\%$. This k value was used without justification, so we use the following R code to calculate the effective number of degrees of freedom and then the applicable t value:

```
> P <- 0.95          # confidence level
> N1 <- 10           # number of data points
> N2 <- 10           # number of data points
>
> # random error variances
> s11 <- 0.000778
> s22 <- 0.00133
>
> # elemental systematic error variances
> bLin11 <- (0.01/2)^2
> bHys11 <- (0.005/2)^2
> bCal11 <- (0.005/2)^2
> bLin22 <- (0.01/2)^2
> bHys22 <- (0.005/2)^2
> bCal22 <- (0.005/2)^2
>
> b11 <- bLin11+bHys11+bCal11
> b22 <- bLin22+bHys22+bCal22
>
> nu_s11 <- N1-1              # number of degrees of freedom
> nu_s22 <- N2-1              # number of degrees of freedom
> nu_bLin11 <- 8             # number of degrees of freedom
> nu_bHys11 <- 8             # number of degrees of freedom
> nu_bCal11 <- 8             # number of degrees of freedom
> nu_bLin22 <- 8             # number of degrees of freedom
> nu_bHys22 <- 8             # number of degrees of freedom
> nu_bCal22 <- 8             # number of degrees of freedom
>
> theta1 <- -1
> theta2 <- 1
> Z <- theta1^2*s11+theta2^2*s22+theta1^2*b11+theta2^2*b22
> rhs <- s11^2/nu_s11+s22^2/nu_s22
> rhs <- rhs+bLin11^2/nu_bLin11+bHys11^2/nu_bHys11+bCal11^2/nu_bCal11
> rhs <- rhs+bLin22^2/nu_bLin22+bHys22^2/nu_bHys22+bCal22^2/nu_bCal22
> nu <- Z^2/rhs           # effective number of degrees of freedom
> t_value <- qt((1+P)/2,nu)   # effective t value
> nu
[1] 18.05287
> t_value
[1] 2.100481
```

The effective number of degrees of freedom is $\nu = 18.05$ and the t value for a 95% confidence level is $t_{\nu,P} = 2.1$, which is about 2. For $\nu > \sim10$, k is close to 2, and for $v \rightarrow \infty$, k approaches a value of 1.96. As a result, it is reasonable to just use $k = 2$ for $P = 95\%$ as long as $v > 10$ (roughly). This criterion is typically satisfied in experiments, and unless otherwise stated, we assume it applies.

6.3 The Kline–McClintock Equation

At this point it is worthwhile to consider the case where all covariance terms are negligible. An expression for u_r^2 for this case can be written as

$$u_r^2 = \theta_1^2 \left(s_1^2 + b_1^2 \right) + \theta_2^2 \left(s_2^2 + b_2^2 \right) + \cdots. \tag{6.44}$$

If we note that Equation 6.45 is the standard uncertainty of x_m,

$$u_m = \sqrt{s_m^2 + b_m^2}, \tag{6.45}$$

we can express Equation 6.44 as

$$u_r^2 = \left(\theta_1 u_1 \right)^2 + \left(\theta_2 u_2 \right)^2 + \cdots = \sum_{m=1}^{L} \left(\theta_m u_m \right)^2. \tag{6.46}$$

The square root of Equation 6.46,

$$u_r = \sqrt{\sum_{m=1}^{L} \left(\theta_m u_m \right)^2}, \tag{6.47}$$

is a form of the Kline–McClintock equation that is sometimes used for error propagation [3]. It is valid when the covariance terms are negligible. Equation 6.47 gives the standard uncertainty in r. An analogous expression can be developed for the standard uncertainty in the *mean* of r:

$$u_{\bar{r}} = \sqrt{\sum_{m=1}^{L} \theta_m^2 \left(\frac{s_m^2}{N} + b_m^2 \right)}. \tag{6.48}$$

Equations 6.47 and 6.48 can be multiplied by a coverage factor k if desired.

6.4 Balance Checks

Certain physical quantities such as mass, energy, charge, and momentum are conserved, and when we write down the equations to describe their conservation, we generally assume a balance exists between different terms, e.g., for a control volume. For example, conservation of mass applied to a control volume can be expressed as

$$\frac{dm_{cv}}{dt} = \sum_{in} \dot{m}_{in} - \sum_{out} \dot{m}_{out}. \qquad (6.49)$$

If the left-hand side of Equation 6.49 is zero, i.e., conditions are steady, mass conservation is then

$$0 = \sum_{in} \dot{m}_{in} - \sum_{out} \dot{m}_{out}. \qquad (6.50)$$

The balance check in this case then consists of equating Equation 6.50 to a variable we call r, so

$$r = \sum_{in} \dot{m}_{in} - \sum_{out} \dot{m}_{out}. \qquad (6.51)$$

The mass flow rates in Equation 6.51 would be measured and would thus have uncertainties. Equation 6.51 can be treated as a data reduction equation and the uncertainty in r can be evaluated. We would consider Equation 6.50 to be applicable if the following condition is satisfied [4]:

$$|\bar{r}| \leq ku_{\bar{r}}. \qquad (6.52)$$

Note that k is a coverage factor that depends on the coverage level selected. Equation 6.52 would also apply to other balance equations.

Problems

P6.1 Show the steps required to go from Equations 6.6 to 6.7.

P6.2 Show the steps required to go from Equations 6.22 to 6.24.

P6.3 A set of calipers is used to measure the sides x and y of a rectangle. Each side is measured 10 times and the resulting data are as
follows:

x (mm)	9.93	9.91	10.14	10.19	10.00	10.01	10.02	10.03	10.11	10.01
y (mm)	19.87	20.07	20.10	19.98	20.21	20.07	19.88	20.08	20.01	19.86

Calculate the mean area of the rectangle as well as its 95% uncertainty interval. The calipers have the following elemental systematic
errors: linearity = 0.01 mm, hysteresis = 0.005 mm, and calibration =
0.005 mm. Assume that the relative uncertainties of the elemental systematic errors are all 10%.

P6.4 The term b_{mn} that appears in Equation 6.32 was asserted to be the sum
of the covariances of the elemental systematic errors that are common
to $\bar{x}_m$ and $\bar{x}_n$. Justify this assertion.

P6.5 Oil flowing through a pipe enters a flowmeter (flowmeter 1, for which
$\dot{m}_1 = 3.97$ kg/s). The oil exiting this flowmeter splits into two flows,
with each flow having its own flowmeter (flowmeters 2 and 3, for
which $\dot{m}_2 = 3.02$ kg/s and $\dot{m}_3 = 1.03$ kg/s). Perform a balance check
on mass conservation at a 95% confidence level assuming that random
errors are negligible. Each flowmeter has a standard systematic error
of 0.02 kg/s from installation. In addition, flowmeters 2 and 3, which
were calibrated using the same standard, have standard systematic
uncertainties (b_{cal}) of 2% of the reading. Flowmeter 1 was calibrated
against a different standard and has a standard systematic uncertainty of 2% of the reading. Note that mass conservation requires that
$\dot{m}_1 = \dot{m}_2 + \dot{m}_3$.

P6.6 We want to use measurements of current (I), voltage drop (V), and
resistance (R) to calculate the power dissipation in a resistor. Each
measured value has a relative uncertainty of 0.05 at a 95% confidence level. Is it better to calculate the power dissipation as (1) IV,
(2) I^2R, or (3) V^2/R? The method you select should yield the lowest
uncertainty.

P6.7 The drag coefficient of a sphere immersed in a fluid is defined as $C_D =
8F/(\pi \rho V^2 d^2)$, where F is the drag force exerted on the sphere, ρ is the
fluid density, V is the fluid–sphere relative velocity, d is the sphere
diameter, and μ is the dynamic viscosity of the fluid. What is the 95%
relative uncertainty in C_D if the standard relative uncertainties in all
other variables are 2%?

P6.8 A horizontal water jet is turned through the angle of $\pi/2$ by a vane.
The horizontal force exerted by the water on the vane is measured
as $F = 120$ N, the water mass flow rate is measured as $\dot{m} = 1.24$ kg/s,

and the jet's (horizontal) velocity just before the vane is measured to be $V = 105$ m/s. Perform a balance check on (horizontal) momentum conservation at a 95% confidence level. The water mass flow rate is measured with a flowmeter that has a standard systematic error of 0.02 kg/s from installation as well as a standard systematic uncertainty of 2% of the reading. Flowmeter random errors are negligible. The velocity and force measurements both have relative uncertainties of 2% (at a 95% confidence level). Assume that momentum conservation requires $F = \dot{m}V$.

P6.9 The Nusselt number of a sphere immersed in a fluid is defined as $\mathrm{Nu} = hD/\lambda$, where h is the convective heat transfer coefficient, D is the sphere diameter, and λ is the thermal conductivity of the fluid. Experiments have indicated that, for a particular condition, $h = 103$ W/(m² K), $D = 25.3$ mm, and $\lambda = 0.105$ W/(m K). The measurements also show that $s_h = 2.5$ W/(m² K), $b_h = 1.1$ W/(m² K), s_D is negligible, $b_D = 0.1$ mm, s_λ is negligible, and $b_\lambda = 0.002$ W/(m K). Calculate the 95% relative uncertainty in Nu.

P6.10 Develop an expression to predict the *minimum* standard uncertainty of the mean for a result calculated using experimental data (assuming that N is very large). Simplify this expression for the situation where all covariances are negligible.

References

1. JCGM 100:2008, Evaluation of measurement data—Guide to the expression of uncertainty in measurement, GUM 1995 with minor corrections, International Bureau of Weight and Measures (BIPM), Sérres, France, 2008.
2. ASME PTC 19.1-2013, Test uncertainty—Performance test codes, The American Society of Mechanical Engineers, New York, 2013.
3. S. J. Kline and F. A. McClintock, Describing the uncertainties in single sample experiments, *Mechanical Engineering*, 75, 3–8 (1953).
4. H. W. Coleman and W. G. Steele, *Experimentation, Validation, and Uncertainty Analysis for Engineers*, 3rd edn., Wiley, Hoboken, NJ, 2009.

7

Taylor Series Uncertainty of a Linear Regression Curve Fit

We previously considered a few techniques for fitting curves to data: (1) linear regression, (2) nonlinear regression, and (3) kernel smoothing. These methods allowed us to draw a best-fit line through a set of data. We will now address the uncertainties of best-fit lines for cases where linear regression is used. The focus of this chapter will be on using a Taylor series approach. Monte Carlo methods [1], which can readily be used for nonlinear regression and kernel smoothing approaches, will be discussed in the next chapter.

7.1 Curve-Fit Expressions

We consider curve-fit expressions of the form

$$y_c = a_0 f_0(x) + a_1 f_1(x) + a_2 f_2(x) + \cdots + a_n f_n(x), \tag{7.1}$$

where $f_0(x), f_1(x),\ldots$ could be nonlinear functions of x, the subscript c on y_c indicates a curve fit, and the coefficients $a_0, a_1,\ldots$ appear linearly.

With the Taylor series approach we note that, because the coefficients are functions of the data points,

$$a_i = a_i(x_1, x_2,\ldots, x_N, y_1, y_2,\ldots, y_N), \tag{7.2}$$

the curve fit itself is a function of the data points and x:

$$y_c = y_c(x_1, x_2,\ldots, x_N, y_1, y_2,\ldots, y_N, x). \tag{7.3}$$

To calculate the variance of y_c, we will propagate errors into the curve fit using Taylor series expansion techniques [3], leading to

$$u_{y_c}^2 = u_{xx} + u_{yy} + 2u_{xy}, \tag{7.4}$$

where u_{xx}, u_{yy}, and u_{xy} are defined as follows:

$$u_{xx} = \sum_{m=1}^{N}\sum_{n=1}^{N}\left(\frac{\partial y_c}{\partial x_m}\right)\left(\frac{\partial y_c}{\partial x_n}\right)C(x_m,x_n),$$ (7.5)

$$u_{yy} = \sum_{m=1}^{N}\sum_{n=1}^{N}\left(\frac{\partial y_c}{\partial y_m}\right)\left(\frac{\partial y_c}{\partial y_n}\right)C(y_m,y_n),$$ (7.6)

and

$$u_{xy} = \sum_{m=1}^{N}\sum_{n=1}^{N}\left(\frac{\partial y_c}{\partial x_m}\right)\left(\frac{\partial y_c}{\partial y_n}\right)C(x_m,y_n).$$ (7.7)

The variables x_i and y_i are treated as random variables and are given by

$$x_i = \langle x_i \rangle + \alpha_{x_i} + \left(\varepsilon_{x_{i,1}} + \varepsilon_{x_{i,2}} + \cdots\right) + \left(\beta_{x_{i,1}} + \beta_{x_{i,2}} + \cdots\right)$$ (7.8)

and

$$y_i = \langle y_i \rangle + \alpha_{y_i} + \left(\varepsilon_{y_{i,1}} + \varepsilon_{y_{i,2}} + \cdots\right) + \left(\beta_{y_{i,1}} + \beta_{y_{i,2}} + \cdots\right),$$ (7.9)

where we allow for multiple random and systematic elemental errors with the ε and β terms as well as variations in x_i and y_i with the α terms.

For brevity we write Equations 7.8 and 7.9 as

$$x_i = \langle x_i \rangle + \alpha_{x_i} + \varepsilon_{x_i} + \beta_{x_i}$$ (7.10)

and

$$y_i = \langle y_i \rangle + \alpha_{y_i} + \varepsilon_{y_i} + \beta_{y_i},$$ (7.11)

where $\varepsilon_{x_i} = \varepsilon_{x_{i,1}} + \varepsilon_{x_{i,2}} + \cdots$, $\beta_{x_i} = \beta_{x_{i,1}} + \beta_{x_{i,2}} + \cdots$, $\varepsilon_{y_i} = \varepsilon_{y_{i,1}} + \varepsilon_{y_{i,2}} + \cdots$, and $\beta_{y_i} = \beta_{y_{i,1}} + \beta_{y_{i,2}} + \cdots$. For later analyses it is worthwhile to recall that the covariance of two linear sums of random variables $P = p_1 + p_2 + \cdots + p_L$ and $Q = q_1 + q_2 + \cdots + q_M$ is given by

$$C(P,Q) = C(p_1 + p_2 + \cdots + p_L, q_1 + q_2 + \cdots + q_M) = \sum_{i=1}^{L}\sum_{j=1}^{M}C(p_i,q_j).$$ (7.12)

Also note that $C(p_i,q_j) = 0$ if p_i and q_j are uncorrelated, that $C(p_i,p_i) = V(p_i)$, and that $C(p_i,q_j) = \rho_{p_i q_j}V(p_i)V(q_j)$.

The derivatives in Equations 7.4 through 7.7 are evaluated using

$$\frac{\partial y_c}{\partial x_m} = \frac{\partial a_0}{\partial x_m} f_0(x) + \frac{\partial a_1}{\partial x_m} f_1(x) + \frac{\partial a_2}{\partial x_m} f_2(x) + \cdots + \frac{\partial a_n}{\partial x_m} f_n(x) \qquad (7.13)$$

and

$$\frac{\partial y_c}{\partial y_m} = \frac{\partial a_0}{\partial y_m} f_0(x) + \frac{\partial a_1}{\partial y_m} f_1(x) + \frac{\partial a_2}{\partial y_m} f_2(x) + \cdots + \frac{\partial a_n}{\partial y_m} f_n(x), \qquad (7.14)$$

where we assume that there is no uncertainty in the value of x we use as an input.

Consider as an example the straight-line curve fit of

$$y_c = a_0 + a_1 x. \qquad (7.15)$$

The coefficients a_0 and a_1 in Equation 7.15 can be found analytically [2,3] to be

$$a_0 = \frac{\left(\sum_{i=1}^{N} x_i^2\right)\left(\sum_{i=1}^{N} y_i\right) - \left(\sum_{i=1}^{N} x_i\right)\left(\sum_{i=1}^{N} x_i y_i\right)}{N \sum_{i=1}^{N} x_i^2 - \left(\sum_{i=1}^{N} x_i\right)^2} \qquad (7.16)$$

and

$$a_1 = \frac{N\left(\sum_{i=1}^{N} x_i y_i\right) - \left(\sum_{i=1}^{N} x_i\right)\left(\sum_{i=1}^{N} y_i\right)}{N\left(\sum_{i=1}^{N} x_i^2\right) - \left(\sum_{i=1}^{N} x_i\right)^2}. \qquad (7.17)$$

Equations 7.13 and 7.14 reduce to

$$\frac{\partial y_c}{\partial x_m} = \frac{\partial a_0}{\partial x_m} + \frac{\partial a_1}{\partial x_m} x \qquad (7.18)$$

and

$$\frac{\partial y_c}{\partial y_m} = \frac{\partial a_0}{\partial y_m} + \frac{\partial a_1}{\partial y_m} x, \qquad (7.19)$$

where the derivatives that appear on the right-hand side of Equations 7.18 and 7.19 are given by

$$\frac{\partial a_0}{\partial x_m} = \frac{2x_m \sum_{i=1}^{N} y_i - \sum_{i=1}^{N} x_i y_i - y_m \left(\sum_{i=1}^{N} x_i \right) - 2Na_0 x_m + 2a_0 \sum_{i=1}^{N} x_i}{N \sum_{i=1}^{N} x_i^2 - \left(\sum_{i=1}^{N} x_i \right)^2}, \quad (7.20)$$

$$\frac{\partial a_1}{\partial x_m} = \frac{N y_m - \sum_{i=1}^{N} y_i - 2Na_1 x_m + 2a_1 \sum_{i=1}^{N} x_i}{N \sum_{i=1}^{N} x_i^2 - \left(\sum_{i=1}^{N} x_i \right)^2}, \quad (7.21)$$

$$\frac{\partial a_0}{\partial y_m} = \frac{\left(\sum_{i=1}^{N} x_i^2 \right) - x_m \left(\sum_{i=1}^{N} x_i \right)}{N \sum_{i=1}^{N} x_i^2 - \left(\sum_{i=1}^{N} x_i \right)^2}, \quad (7.22)$$

and

$$\frac{\partial a_1}{\partial y_m} = \frac{N x_m - \sum_{i=1}^{N} x_i}{N \sum_{i=1}^{N} x_i^2 - \left(\sum_{i=1}^{N} x_i \right)^2}. \quad (7.23)$$

We will now use these expressions to assess the uncertainty in Equation 7.15 for several cases presented next.

7.2 Cases to Consider

7.2.1 Case 1: No Errors in x_i and No Correlations

Case 1 is the situation in which each x_i has zero variance, there are no systematic errors in y_i, and there are no correlations.

For this case, Equations 7.10 and 7.11 reduce to

$$x_i = \langle x_i \rangle \quad (7.24)$$

and

$$y_i = \langle y_i \rangle + \alpha_{y_i} + \varepsilon_{y_i}. \quad (7.25)$$

The term $C(x_m,x_n)$ in Equation 7.5 is zero because all of the x_i terms have zero variance, yielding $u_{xx} = 0$. Similarly, the term $C(x_m,y_n)$ in Equation 7.7 is zero because there are no correlations and thus $u_{xy} = 0$. The term $C(y_m,y_n)$ in Equation 7.6 is zero unless $m = n$, in which case $C(y_m,y_m) = V_y = V(\alpha_{y_i}) + V(\varepsilon_{y_i})$. Equation 7.4 then can be written as

$$u_{y_c}^2 = u_{yy} = V_y \sum_{m=1}^{N} \left(\frac{\partial y_c}{\partial y_m} \right)^2 . \tag{7.26}$$

In practice, the sum $V(\alpha_{y_i}) + V(\varepsilon_{y_i})$ is typically estimated as the sum of the squares of the residuals divided by the number of degrees of freedom, $N - 2$:

$$V\left(\alpha_{y_i}\right) + V\left(\varepsilon_{y_i}\right) = \frac{1}{N-2} \sum_{i=1}^{N} \left[y_i - y_c\left(x_i\right) \right]^2 . \tag{7.27}$$

The sample R code for this case is as follows:

```
> # define the data and parameters
> N <- 20
> x <- seq(from=0,to=10,length.out=N)
> set.seed(100)
> y <- x+rnorm(n=N,mean=0,sd=0.2)
> P <- 0.95                        # confidence level
> dof <- N-2                       # degrees of freedom
>
> # x values for the curve-fit plot
> M <- 100
> xplot <- seq(from=min(x),to=max(x),length.out=M)
>
> # calculate the curve fit
> a0 <- (sum(x^2)*sum(y)-sum(x)*sum(x*y))/(N*sum(x^2)-sum(x)^2)
> a1 <- (N*sum(x*y)-sum(x)*sum(y))/(N*sum(x^2)-sum(x)^2)
> yc <- a0+a1*x
> ycplot <- a0+a1*xplot
>
> # evaluate the residuals
> E <- y-yc
>
> # evaluate Vy
> Vy <- sum(E*E)/dof
>
> # evaluate uyy at each xplot value
> uyy <- numeric(M)
> for (m in 1:M) {
+       da0dym <- (sum(x^2)-x*sum(x))/(N*sum(x^2)-sum(x)^2)
+       da1dym <- (N*x-sum(x))/(N*sum(x^2)-sum(x)^2)
+       dycdym <- da0dym+da1dym*xplot[m]
+       uyy[m] <- Vy*sum(dycdym^2)
+ }
>
```

```
> # evaluate uyc at each xplot value
> uyc <- sqrt(uyy)
>
> # plot the data and curve fit
> plot(x,y)                          # original data
> lines(xplot,ycplot)                # best-fit line
>
> # plot the curve-fit uncertainty
> quartz()
> k <- qt((P+1)/2,dof)                          # coverage factor
> plot(xplot,k*uyc,type='l',ylim=c(0.05,0.15))  # k*uyc plot
```

We generate a set of simulated data by adding normal noise to a straight line. Figure 7.1 shows the resulting curve fit and the original data. Figure 7.2 shows the 95% uncertainty in the curve fit as a function of x. The uncertainties are larger near the smallest and largest x values, which is typically the case.

7.2.2 Case 2: Random Errors Only

Case 2 is the situation in which there are random variations and errors in both x_i and y_i but no systematic errors or correlations.

Here, we assume that all of the x_i values share the same random variations and elemental random errors. Similarly, the y_i values are assumed to share random variations and elemental random errors (but that are different than

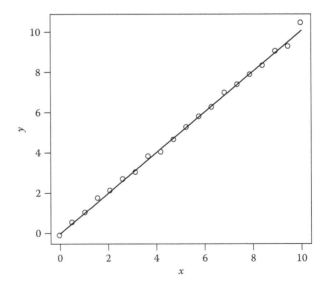

FIGURE 7.1
Plot of data (circles) and a best-fit straight line (case 1: no errors in x_i).

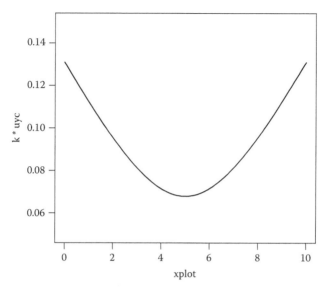

FIGURE 7.2
Uncertainties ($P = 0.95$) for the curve fit in Figure 7.1.

for x_i). We assume that there are no correlations and that systematic errors are negligible. Equations 7.10 and 7.11 then become

$$x_i = \langle x_i \rangle + \alpha_{x_i} + \varepsilon_{x_i} \tag{7.28}$$

and

$$y_i = \langle y_i \rangle + \alpha_{y_i} + \varepsilon_{y_i}. \tag{7.29}$$

Equation 7.4 then can be written as

$$u_{y_c}^2 = V_x \sum_{m=1}^{N} \left(\frac{\partial y_c}{\partial x_m} \right)^2 + V_y \sum_{m=1}^{N} \left(\frac{\partial y_c}{\partial y_m} \right)^2, \tag{7.30}$$

where $V_x = V(\alpha_{x_i}) + V(\varepsilon_{x_i})$ and $V_y = V(\alpha_{y_i}) + V(\varepsilon_{y_i})$. The first term on the right-hand side of Equation 7.32 is u_{xx}, the second term is u_{yy}, and $u_{xy} = 0$.

The R code for this example is as follows:

```
> # define the data and parameters
> N <- 20
> x <- seq(from=0,to=10,length.out=N)
> Vx <- 0.05^2
> set.seed(100)
> y <- 1+2*x+rnorm(n=N,mean=0,sd=0.2)
> P <- 0.95                          # confidence level
> dof <- N-2                         # degrees of freedom
>
> # x values for the curve-fit plot
> M <- 100
> xplot <- seq(from=min(x),to=max(x),length.out=M)
>
```

```
> # calculate the curve fit
> a0 <- (sum(x^2)*sum(y)-sum(x)*sum(x*y))/(N*sum(x^2)-sum(x)^2)
> a1 <- (N*sum(x*y)-sum(x)*sum(y))/(N*sum(x^2)-sum(x)^2)
> yc <- a0+a1*x
> ycplot <- a0+a1*xplot
>
> # evaluate the residuals
> E <- y-yc
>
> # evaluate Vy
> Vy <- sum(E*E)/dof
>
> # evaluate uxx at each xplot value
> uxx <- numeric(M)
> for (m in 1:M) {
+       da0dxm <- (2*x*sum(y)-sum(x*y)-y*sum(x)-a0*(2*N*x-2*sum(x)))/
          (N*sum(x^2)-sum(x)^2)
+       da1dxm <- (N*y-sum(y)-a1*(2*N*x-2*sum(x)))/(N*sum(x^2)-sum(x)^2)
+       dycdxm <- da0dxm+da1dxm*xplot[m]
+       uxx[m] <- Vx*sum(dycdxm^2)
+ }
>
> # evaluate uyy at each xplot value
> uyy <- numeric(M)
> for (m in 1:M) {
+       da0dym <- (sum(x^2)-x*sum(x))/(N*sum(x^2)-sum(x)^2)
+       da1dym <- (N*x-sum(x))/(N*sum(x^2)-sum(x)^2)
+       dycdym <- da0dym+da1dym*xplot[m]
+       uyy[m] <- Vy*sum(dycdym^2)
+ }
>
> # evaluate uyc at each xplot value
> uyc <- sqrt(uxx+uyy)                      # includes Vx
> uyc0 <- sqrt(uyy)                         # neglects Vx
>
> # plot the data and curve fit
> plot(x,y)                                 # original data
> lines(xplot,ycplot)
>
> # plot the curve-fit uncertainty
> quartz()
> k <- qt((P+1)/2,dof)                      # coverage factor
> plot(xplot,k*uyc,type='l',ylim=c(0.05,0.15))  # k*uyc plot
> lines(xplot,k*uyc0,lty='dotted')         # k*uyc0 plot
```

The values for V_x are specified in the code; i.e., we are assuming that we have evaluated V_x using a prior analysis. The variance V_y is estimated with Equation 7.27.

Figure 7.3 shows the original data as circles and the best-fit line as the straight solid line. Figure 7.4 shows plots of the 95% uncertainty in y_c for cases where V_x and V_y are both included (solid line) and where V_x is neglected (dotted line). There is a noticeable difference between these curves, indicating that for this example it is important to account for V_x.

7.2.3 Case 3: Random and Systematic Errors

Case 3 is the situation in which there are random variations and errors in both x_i and y_i. There is a systematic error common to all of the y_i values.

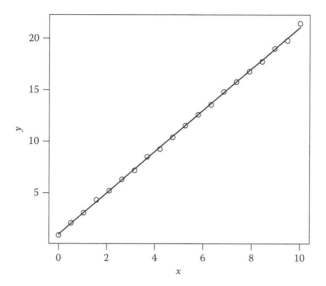

FIGURE 7.3
Plot of data (circles) and a best-fit straight line (case 2: random errors in x_i and y_i).

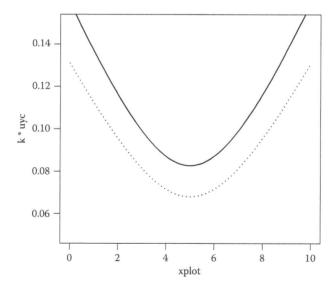

FIGURE 7.4
Uncertainties ($P = 0.95$) for the curve fit in Figure 7.3. The dotted line is for when V_x is neglected and the solid line includes V_x.

Here, we assume that all of the x_i values share the same random variations and elemental random errors. Similarly, the y_i values are assumed to share random variations and elemental random errors (but that are different than for x_i). The y_i values have a common systematic error β_y. Equations 7.10 and 7.11 then reduce to

$$x_i = \langle x_i \rangle + \alpha_{x_i} + \varepsilon_{x_i} \qquad (7.31)$$

and

$$y_i = \langle y_i \rangle + \alpha_{y_i} + \varepsilon_{y_i} + \beta_y. \qquad (7.32)$$

Equation 7.4 then can be written as

$$u_{y_c}^2 = V_x \sum_{m=1}^{N}\left(\frac{\partial y_c}{\partial x_m}\right)^2 + V_y \sum_{m=1}^{N}\left(\frac{\partial y_c}{\partial y_m}\right)^2 + b_y^2\sum_{m=1}^{N}\sum_{n=1}^{N}\left(\frac{\partial y_c}{\partial y_m}\right)\left(\frac{\partial y_c}{\partial y_n}\right), \qquad (7.33)$$

where $V_x = V(\alpha_{x_i}) + V(\varepsilon_{x_i})$, $C(y_m, y_n) = b_y^2$ if $m \neq n$, $C(y_m, y_n) = V_y + b_y^2$ if $m = n$, $V_y = V(\alpha_{y_i}) + V(\varepsilon_{y_i})$, and $b_y^2 = C(\beta_y, \beta_y)$ is the variance of β_y.

The first term on the right-hand side of Equation 7.33 is u_{xx}; u_{yy} is the sum of the second and third terms, and $u_{xy} = 0$. It can be shown that

$$\sum_{m=1}^{N}\sum_{n=1}^{N}\left(\frac{\partial y_c}{\partial y_m}\right)\left(\frac{\partial y_c}{\partial y_n}\right) = 1 \qquad (7.34)$$

holds for Equation 7.15 such that we can write

$$u_{y_c}^2 = V_x \sum_{m=1}^{N}\left(\frac{\partial y_c}{\partial x_m}\right)^2 + V_y \sum_{m=1}^{N}\left(\frac{\partial y_c}{\partial y_m}\right)^2 + b_y^2. \qquad (7.35)$$

The R code for this example is as follows:

```
> # define the data and parameters
> N <- 20
> x <- seq(from=0,to=10,length.out=N)
> Vx <- 0.05^2
> by <- 0.05
> set.seed(200)
> y <- 1-3*x+rnorm(n=N,mean=0,sd=0.5)
> P <- 0.95                              # confidence level
> dof <- N-2                             # degrees of freedom
>
> # x values for the curve-fit plot
> M <- 100
> xplot <- seq(from=min(x),to=max(x),length.out=M)
>
> # calculate the curve fit
> a0 <- (sum(x^2)*sum(y)-sum(x)*sum(x*y))/(N*sum(x^2)-sum(x)^2)
> a1 <- (N*sum(x*y)-sum(x)*sum(y))/(N*sum(x^2)-sum(x)^2)
> yc <- a0+a1*x
> ycplot <- a0+a1*xplot
>
```

```
> # evaluate the residuals
> E <- y-yc
>
> # evaluate Vy
> Vy <- sum(E*E)/dof
>
> # evaluate uxx at each xplot value
> uxx <- numeric(M)
> for (m in 1:M) {
+       da0dxm <- (2*x*sum(y)-sum(x*y)-y*sum(x)-a0*(2*N*x-2*sum(x)))/
          (N*sum(x^2)-sum(x)^2)
+       da1dxm <- (N*y-sum(y)-a1*(2*N*x-2*sum(x)))/(N*sum(x^2)-sum(x)^2)
+       dycdxm <- da0dxm+da1dxm*xplot[m]
+       uxx[m] <- Vx*sum(dycdxm^2)
+ }
>
> # evaluate uyy at each xplot value
> uyy <- numeric(M)
> for (m in 1:M) {
+       da0dym <- (sum(x^2)-x*sum(x))/(N*sum(x^2)-sum(x)^2)
+       da1dym <- (N*x-sum(x))/(N*sum(x^2)-sum(x)^2)
+       dycdym <- da0dym+da1dym*xplot[m]
+       uyy[m] <- Vy*sum(dycdym^2)
+ }
> uyy <- uyy+by^2
>
> # evaluate uyc at each xplot value
> uyc <- sqrt(uxx+uyy)                     # includes Vx and by
> uyc0 <- sqrt(uyy)                        # neglects Vx and by
>
> # plot the data and curve fit
> plot(x,y)                                # original data
> lines(xplot,ycplot)
>
> # plot the curve-fit uncertainty
> quartz()
> k <- qt((P+1)/2,dof)                     # coverage factor
> plot(xplot,k*uyc,type='l',ylim=c(0.2,0.4))  # k*uyc plot
> lines(xplot,k*uyc0,lty='dotted')         # k*uyc0 plot
```

The values for V_x and b_y are specified in the code; i.e., we are assuming that we have evaluated these variables using a prior analysis. The variance V_y is estimated with Equation 7.27.

Figure 7.5 shows the original data as circles and the best-fit line as the straight solid line. Figure 7.6 shows plots of the 95% uncertainty in y_c for cases where V_x, V_y, and b_y are included (solid line) and where V_x and b_y are neglected (dotted line). There is a noticeable difference between these curves, indicating that for this example it is important to account for V_x and b_y.

7.3 General Linear Regression Theory

To extend the theory in the previous section to more general situations, it is advantageous to express linear regression problems in matrix form. This allows us to derive general relations in a compact form that is amenable to programming in computer languages that are vectorized, such as R.

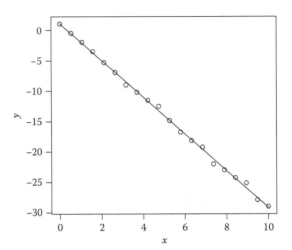

FIGURE 7.5
Plot of data (circles) and a best-fit straight line (case 3: random and systematic errors).

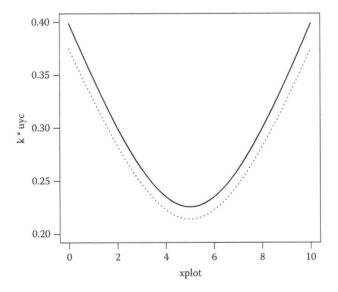

FIGURE 7.6
Uncertainties ($P = 0.95$) for the curve fit in Figure 7.5. The dotted line is for when V_x and b_y are neglected and the solid line includes these terms.

We begin with a curve-fit expression of the form in Equation 7.1. The residual E_i is the difference between data point y_i and the predicted value at x_i:

$$E_i = y_i - y_c(x_i). \tag{7.36}$$

Equation 7.36 can be written in matrix form as

$$E = Y - Y_c, \tag{7.37}$$

where the column vectors E, Y, and Y_c are defined as follows:

$$E = \begin{bmatrix} E_1 \\ E_2 \\ \vdots \\ E_N \end{bmatrix}, \tag{7.38}$$

$$Y = \begin{bmatrix} y_1 \\ y_2 \\ \vdots \\ y_N \end{bmatrix}, \tag{7.39}$$

and

$$Y_c = \begin{bmatrix} y_c(x_1) \\ y_c(x_2) \\ \vdots \\ y_c(x_N) \end{bmatrix}. \tag{7.40}$$

The sum of the squares of the residuals can be expressed as

$$\sum_{i=1}^{N} E_i^2 = E^T E \tag{7.41}$$

such that

$$E^T E = Y^T Y - 2Y_c^T Y + Y_c^T Y_c \tag{7.42}$$

holds when Equation 7.37 is employed (and note that the superscript T notation denotes transpose).

Now define the matrix F as

$$F = \begin{bmatrix} f_0(x_1) & f_1(x_1) & \cdots & f_n(x_1) \\ f_0(x_2) & f_1(x_2) & \cdots & f_n(x_2) \\ \vdots & \vdots & & \vdots \\ f_0(x_N) & f_1(x_N) & \cdots & f_n(x_N) \end{bmatrix} \tag{7.43}$$

such that

$$Y_c = FA \tag{7.44}$$

holds. The column vector A in Equation 7.44, which has the regression coefficients as its elements, is defined as

$$A = \begin{bmatrix} a_0 \\ a_1 \\ \vdots \\ a_n \end{bmatrix}. \tag{7.45}$$

Inserting Equation 7.44 into Equation 7.42 yields

$$E^T E = Y^T Y - 2A^T F^T Y + A^T F^T F A. \tag{7.46}$$

We now minimize the sum of the squares of the residuals by applying

$$\frac{\partial \left(E^T E \right)}{\partial a_i} = -2F^T Y + 2F^T F A = 0, \tag{7.47}$$

leading to

$$A = GY, \tag{7.48}$$

where the matrix G in Equation 7.48 is given by

$$G = \left(F^T F \right)^{-1} F^T. \tag{7.49}$$

The matrix G depends only on the x_i values and the column vector Y only on the y_i values. A predicted value $y_c(x)$ is given by

$$y_c(x) = A^T H, \tag{7.50}$$

where H is the column vector

$$H = \begin{bmatrix} f_0(x) \\ f_1(x) \\ \vdots \\ f_n(x) \end{bmatrix}. \tag{7.51}$$

The predicted y_c values at the x_i data points are given by Equation 7.44.

Equation 7.4 is still used to calculate the variance of y_c with this more general theory. However, to use Equation 7.4, we need to be able to evaluate the partial derivatives $\partial y_c/\partial x_m$ and $\partial y_c/\partial y_m$. The derivatives with respect to x_m are evaluated with Equation 7.13, which can be expressed in matrix form. To do so, we first define a column vector K that contains the $\partial y_c/\partial x_m$ values:

$$K = \begin{bmatrix} \dfrac{\partial y_c}{\partial x_1} \\[2mm] \dfrac{\partial y_c}{\partial x_2} \\[1mm] \vdots \\[1mm] \dfrac{\partial y_c}{\partial x_N} \end{bmatrix}. \qquad (7.52)$$

Following this, we relate the elements of K to partial derivatives of A as

$$K = \begin{bmatrix} H^T \dfrac{\partial A}{\partial x_1} \\[2mm] H^T \dfrac{\partial A}{\partial x_2} \\[1mm] \vdots \\[1mm] H^T \dfrac{\partial A}{\partial x_N} \end{bmatrix}. \qquad (7.53)$$

The derivatives of A that appear in Equation 7.53 can be evaluated by differentiating Equation 7.47 with respect to x_m, yielding

$$\frac{\partial A}{\partial x_m} = \left(F^T F\right)^{-1} \left(\frac{\partial F}{\partial x_m}\right)^T Y - \left(F^T F\right)^{-1} \left(\left(\frac{\partial F}{\partial x_m}\right)^T F + F^T \left(\frac{\partial F}{\partial x_m}\right)\right) A. \qquad (7.54)$$

The $\partial y_c/\partial y_m$ derivatives are evaluated using Equation 7.14. From Equation 7.48 it follows that the derivatives on the right-hand side of Equation 7.14 are given by

$$\frac{\partial a_j}{\partial y_m} = G_{jm}, \qquad (7.55)$$

where G_{jm} is the element in row j and column m of the matrix G defined in Equation 7.49.

We also define the column vector L as

$$
L = \begin{bmatrix} \dfrac{\partial y_c}{\partial y_1} \\[2mm] \dfrac{\partial y_c}{\partial y_2} \\[2mm] \vdots \\[2mm] \dfrac{\partial y_c}{\partial y_N} \end{bmatrix}.
$$

(7.56)

Note that L is related to the matrix G and the column vector H via

$$
L = G^T H.
$$

(7.57)

Here is an example where we are fitting a curve of the form

$$
y_c = a_0 + a_1 x + a_2 \ln(1+x)
$$

(7.58)

to a set of data. Note that $f_0(x) = 1, f_1(x) = x$, and $f_2(x) = \ln(1 + x)$. For this example, we assume that there are random variations and errors in both x_i and y_i but no systematic errors or correlations. Equations 7.10 and 7.11 then become

$$
x_i = \langle x_i \rangle + \alpha_{x_i} + \varepsilon_{x_i}
$$

(7.59)

and

$$
y_i = \langle y_i \rangle + \alpha_{y_i} + \varepsilon_{y_i}.
$$

(7.60)

Equation 7.4 then can be written as

$$
u_{y_c}^2 = V_x \sum_{m=1}^{N} \left(\frac{\partial y_c}{\partial x_m} \right)^2 + V_y \sum_{m=1}^{N} \left(\frac{\partial y_c}{\partial y_m} \right)^2,
$$

(7.61)

where $V_x = V(\alpha_{x_i}) + V(\varepsilon_{x_i})$ and $V_y = V(\alpha_{y_i}) + V(\varepsilon_{y_i})$. Equation 7.61 can be expressed in matrix form as

$$
u_{y_c}^2 = V_x K^T K + V_y L^T L.
$$

(7.62)

The R code for this example is as follows:

```
> # define the data and parameters
> Vx <- 0.05^2
> x <- seq(from=1,to=10,length.out=20)
> y <- 1+2*x+3*log(1+x)+rnorm(n=length(x),mean=0,sd=0.5)
> N <- length(x)
```

```
> P <- 0.95
> Y <- cbind(y)
> N <- length(x)
>
> # calculate the F matrix
> f0 <- rep(1,N)
> f1 <- x
> f2 <- log(1+x)
> F <- cbind(f0,f1,f2)
> n <- length(F[1,])-1
> dof <- N-1-n
>
> # calculate the G matrix
> G <- solve((t(F)%*%F))%*%t(F)
>
> # calculate the A matrix
> A <- G%*%Y
> a0 <- A[1]
> a1 <- A[2]
> a2 <- A[3]
>
> # evaluate the residuals and variance in y
> yc <- a0+a1*x+a2*log(1+x)
> Yc <- cbind(yc)
> E <- (Y-Yc)
> Vy <- (t(E)%*%E)/dof
>
> # evaluate local values of KTK where x_local = x[m]
> K <- numeric(N)
> KTK <- numeric(N)
> for (j in 1:N) {
+     for (m in 1:N) {
+             H <- cbind(F[j,])
+             # the next four lines set up the dF/dxm matrix
+                     df0dxm <- rep(0,N)
+                     df1dxm <- rep(0,N)
+                     df1dxm[m] <- 1
+                     df2dxm <- rep(0,N)
+                     df2dxm[m] <- 1/(1+x[m])
+                     dFdxm <- cbind(df0dxm,df1dxm,df2dxm)
+             dAdxm <- t(dFdxm)%*%Y
+             dAdxm <- dAdxm-(t(dFdxm)%*%F+t(F)%*%dFdxm)%*%A
+             dAdxm <- solve(t(F)%*%F)%*%dAdxm
+             K[m] <- t(H)%*%dAdxm
+     }
+     KTK[j] <- sum(K*K)
+ }
>
> # evaluate local values of LTL where x_local = x[m]
> LTL <- numeric(N)
```

```
> for (m in 1:N) {
+      H <- cbind(F[m,])
+      L <- t(G)%*%H
+      LTL[m] <- t(L)%*%L
+ }
>
> # evaluate local uncertainties in yc where x_local = x[m]
> uyc <- sqrt(Vx*KTK+Vy*LTL)            # local uyc including Vx
> uyc0 <- sqrt(Vy*LTL)                  # local uyc excluding Vx
>
> # plot the data and best-fit line
> k <- qt((P+1)/2,dof)
> plot(x,y)                             # original data
> lines(x,yc)                           # best-fit line
>
> # plot the curve-fit uncertainty
> quartz()
> plot(x,k*uyc,type='l')                # k*uyc plot including Vx
> lines(x,k*uyc0,lty='dotted')          # k*uyc plot excluding Vx
```

The values for V_x are specified in the code; i.e., we are assuming that we have evaluated V_x using a prior analysis. The variance V_y is estimated with

$$V_y = \frac{E^T E}{\nu},$$ (7.63)

where the number of degrees of freedom, ν, in Equation 7.63 is obtained from

$$\nu = N - 1 - n.$$ (7.64)

The matrix F in this example is given by

$$F = \begin{bmatrix} 1 & x_1 & \ln(1+x_1) \\ 1 & x_2 & \ln(1+x_2) \\ \vdots & \vdots & \vdots \\ 1 & x_N & \ln(1+x_N) \end{bmatrix}.$$ (7.65)

The matrix $\partial F / \partial x_m$ will have all zeros in the first column and all other elements will be zero except for the second and third columns in row m, which will have the entries 1 and $1/(1 + x_m)$. For example, $\partial F / \partial x_2$ would be written as

$$\frac{\partial F}{\partial x_2} = \begin{bmatrix} 0 & 0 & 0 \\ 0 & 1 & \dfrac{1}{1+x_2} \\ 0 & 0 & 0 \\ \vdots & \vdots & \vdots \\ 0 & 0 & 0 \end{bmatrix}.$$ (7.66)

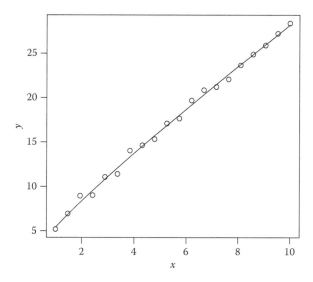

FIGURE 7.7
Plot of data (circles) and a best-fit line.

Figure 7.7 shows the original data as circles and the best-fit line as the solid line. Figure 7.8 shows plots of the 95% uncertainty in y_c for cases where V_x and V_y are included (solid line) and where V_x is neglected (dotted line). There is a noticeable difference between these curves, indicating that for this example it is important to account for V_x.

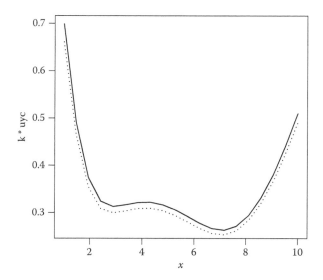

FIGURE 7.8
Uncertainties ($P = 0.95$) for the curve fit in Figure 7.7. The dotted line is for when V_x is neglected and the solid line includes this term.

7.4 Uncertainties in Regression Coefficients

We will now consider briefly the uncertainties in regression coefficients. After considering Equation 7.2, to evaluate the variance of regression coefficient a_i, we write

$$u_{a_i}^2 = \sum_{m=1}^{N}\sum_{n=1}^{N}\left(\frac{\partial a_i}{\partial x_m}\right)\left(\frac{\partial a_i}{\partial x_n}\right)C(x_m,x_n) + \sum_{m=1}^{N}\sum_{n=1}^{N}\left(\frac{\partial a_i}{\partial y_m}\right)\left(\frac{\partial a_i}{\partial y_n}\right)C(y_m,y_n)$$

$$+2\sum_{m=1}^{N}\sum_{n=1}^{N}\left(\frac{\partial a_i}{\partial x_m}\right)\left(\frac{\partial a_i}{\partial y_n}\right)C(x_m,y_n). \tag{7.67}$$

Consider an example where we are fitting a straight line of the form shown in Equation 7.15 to a set of data. Note that $f_0(x) = 1$ and $f_1(x) = x$. For this example, we assume that there are only random variations and errors in y_i. Equations 7.10 and 7.11 then become

$$x_i = \langle x_i \rangle \tag{7.68}$$

and

$$y_i = \langle y_i \rangle + \alpha_{y_i} + \varepsilon_{y_i}. \tag{7.69}$$

Equation 7.67 then can be written as

$$u_{a_i}^2 = V_y \sum_{m=1}^{N}\left(\frac{\partial a_i}{\partial y_m}\right)^2, \tag{7.70}$$

where $V_y = V(\alpha_{y_i}) + V(\varepsilon_{y_i})$.

Using Equations 7.22 and 7.23 in Equation 7.70 yields the expressions

$$u_{a_0} = \left[V_y\left(\frac{1}{N}+\frac{\bar{x}^2}{S_{xx}}\right)\right]^{1/2} \tag{7.71}$$

and

$$u_{a_1} = \left(\frac{V_y}{S_{xx}}\right)^{1/2}, \tag{7.72}$$

where the variable s_{xx} is defined as

$$s_{xx} = \sum_{i=1}^{N} x_i^2 - \frac{1}{N}\left(\sum_{i=1}^{N} x_i\right)^2. \tag{7.73}$$

It is possible to derive similar results for more complex cases. To do so, we need to evaluate the derivatives and covariances as needed.

7.5 Evaluating Uncertainties with Built-in R Functions

R has a built-in capability to plot $P\%$ confidence bands for a curve fit using the `predict()` function as well as to generate $P\%$ confidence intervals for the coefficients. The `predict()` function uses results from `lm()`. A drawback of the `predict()` function is that it only accounts for random errors in y that are estimated from the data used in the fit. If we want to include systematic errors in y as well as random and systematic errors in x, then we need to apply the theory developed earlier in this chapter. Here is an example where we have 10 data points and we are fitting a curve of the form $Y = a_0 + a_1 x + a_2 \ln(1+x)$. The true (noiseless) data are described by $y = 1 + 2x + 3\ln(1-x)$. Here is the R code:

```
> x <- c(0,1,2,3,4,5,6,7,8,9,10)
> y <- c(-0.75,3.37,6.3,10.46,13.70,16.82,19.77,19.19,22.73,26.39,26.88)
> yTrue <- 1+2*x+3*log(1+x)
> N <- length(x)                    # number of data points
> P <- 0.95                         # confidence level
> # open a plot window
> plot(x,y,type="n",xlab="x",ylab="y",xlim=c(0,10),ylim=c(0,28))
> fit <- lm(y~x+I(log(1+x)))                    # generate a fit
> fitPredict <- predict(fit, interval = "confidence")
> lines(x,fitPredict[,"fit"],col="black")       # plot the fit
>
> # plot the original data points
> points(x,y)
>
> # plot the true (noise-free) data
> lines(x,yTrue,col="black",Hy="dashed")
>
> # plot the R confidence band
> lines(x, fitPredict[, "lwr"], lty = "dotted", col="black")
> lines(x, fitPredict[, "upr"], lty = "dotted", col="black")
>
> # print the a0 and a1 coefficients from R
> fit$coefficients
  (Intercept)               x I(log(1 + x))
    -1.230236        1.764123      4.648863
>
> # print the a0 and a1 P% confidence intervals using R
> confint(fit,level=P)
                    2.5 %     97.5 %
(Intercept)     -3.397760  0.9372869
x                1.040218  2.4880280
I(log(1 + x))    1.426627  7.8710985
```

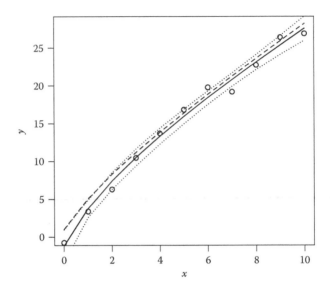

FIGURE 7.9
Confidence band (dotted lines) and a `predict ()` best-fit line (solid line) for a set of data (circles).

Figure 7.9 shows the results from these calculations. The fit (the solid line) follows the data (the circles) fairly well and the true data (the dashed line) fall within or on the 95% confidence band (the dotted lines), though it should be emphasized that in a practical situation we would not actually know the true data.

Problems

P7.1 Consider a curve fit of the form $y_c = a_0 x$. Derive an expression for a_0 and its standard uncertainty. There are no uncertainties in x and the only uncertainty in y is a systematic error.

P7.2 Derive Equations 7.16 and 7.17 by applying Equation 7.48.

P7.3 Verify Equation 7.12.

P7.4 Verify Equation 7.34.

P7.5 Solve the problem in Case 1 (Section 7.2.1) of this chapter using the linear algebra approach described in Section 7.3. Compare your solution with results obtained using the `predict ()` function in R.

P7.6 Solve the problem in Case 2 (Section 7.2.2) of this chapter using the linear algebra approach described in Section 7.3. Compare your solution with results obtained using the `predict ()` function in R.

P7.7 Solve the problem in Case 3 (Section 7.2.3) of this chapter using the linear algebra approach described in Section 7.3. Compare your solution with results obtained using the `predict()` function in R.

P7.8 Consider a curve fit of the form $y_c = a_0 + a_1 x$. Derive expressions for the standard uncertainties in a_0 and a_1 if the x_i values have a common systematic error β_x. Consider all other errors and variations to be negligible.

P7.9 A baseball is thrown into the air. A camera records the baseball's trajectory and measurements yield the following data for the vertical altitude y vs. the horizontal distance traveled, x:

x_i (m)	y_i (m)
0	2
1	11
2	18
3	23
4	26
5	27
6	26
7	23
8	18
9	11

The x_i and y_i measurements are obtained from photographs using a ruler that has a systematic error β. Fit an equation of the form $y_c = a_0 + a_1 x + a_2 x^2$ to these data and evaluate the standard uncertainty in y_c if $V(\beta) = 0.025$ m^2.

P7.10 Consider a curve fit of the form $y_c = a_0 + a_1 x$. Derive an expression for the standard uncertainty in y_c if the x_i values are known exactly and there are no systematic errors. Assume that all of the y_i values have the same variance.

References

1. JCGM 101:2008, Evaluation of measurement data: Supplement 1 to the "Guide to the expression of uncertainty in measurement"—Propagation of distributions using a Monte Carlo method, International Bureau of Weights and Measures (BIPM), Sérres, France, 2008.
2. M. J. Crawley, *The R Book*, 2nd edn., Wiley, West Sussex, U.K., 2013.
3. H. W. Coleman and W. G. Steele, *Experimentation, Validation, and Uncertainty Analysis for Engineers*, 3rd edn., Wiley, Hoboken, NJ, 2009.

8

Monte Carlo Methods

8.1 Overall Monte Carlo Approach

Monte Carlo (MC) methods [1,2] can be viewed in the present context as providing a means for computationally simulating the stochastic nature of errors and their influences on experimental and calculated results. An MC simulation involves generating random values for variables (x_i) of interest, over and over many times, and calculating a result for each set of variables.

The results that are generated are sorted in nondecreasing order [1] to provide a cumulative distribution function (cdf) that can be used to estimate coverage intervals. References [1,2] describe recommended procedures for determining coverage intervals. The quantile() function in R can also be used to determine coverage intervals [3] (and this function can presently calculate coverage intervals in nine different ways). When the number of MC replicates M is large, the coverage factors estimated using quantile() or from [1,2] are generally quite close to each other. We thus use the default method in quantile() here, but the reader should note that there are other approaches that can be used.

Two MC options that we will discuss are (1) the parametric bootstrap and (2) the nonparametric bootstrap. With the parametric bootstrap, you select the distribution(s) that you think represent your data. You use your data to estimate parameters for the distribution(s), and then generate MC replicates. These replicates are then used to calculate the output. The statistics of interest would be calculated after a sufficient number of replicates have been generated.

With a nonparametric bootstrap, you resample, with replacement, from your original N-element data set. Each resampled data set has N elements. The resampled data sets are used to calculate the r values. Statistics of interest would be calculated after a sufficient number of replicates have been generated. An advantage of the nonparametric bootstrapping technique is that one does not need to make any assumptions about the distribution of the population from which the original N data points were drawn. As a result, N should be large enough so that the sample adequately represents the population.

The bootstrapping technique works best when N is large relative to unity, and it has been suggested that N should be at least as large as 20 but that having N as small as 10 may be adequate [4].

Two topics of importance to MC approaches are random number generation and random sampling. These are described next.

8.2 Random Number Generation

The generation of random numbers is important in statistics and data analysis because we can use such numbers to simulate phenomena such as random errors. We need to make sure that the random numbers are drawn from an appropriate pool of numbers, however. This means that it is important to specify the probability distribution. For analysis of experiments, the most common distributions are (1) normal, (2) uniform (rectangular), and (3) triangular. The normal distribution is used for both random and systematic errors, and uniform and triangular distributions can appear when systematic errors are considered.

Here are the R commands for generating random numbers from the normal, uniform, and triangular distributions:

```
> A <- rnorm(n=1000000,mean=-7,sd=1)       # normal distribution
> B <- runif(n=1000000,min=-2,max=2)       # uniform distribution
> C <- rtriangle(n=1000000,a=4,b=8,c=6)    # triangular distribution
```

In each case, we generate one million random numbers subject to various parameters. A histogram (density plot) can be used to show the shape of each distribution of numbers. For simplicity, we show them all on the same plot (Figure 8.1), which we can do because the data sets A, B, and C do not overlap. Here is the R code:

```
> D <- c(A,B,C)                       # combine the distributions
> hist(D,n=200,prob=TRUE,main="")     # generate the histogram
```

The random number generator algorithms in R actually generate pseudorandom numbers, meaning that you will always get the same number sequence if you start with the same number; i.e., you use the same "seed" each time for the random number generator. This can be done with the set.seed() function, but if you do not use this function, then R selects its own starting point, which would generally be different every time random numbers are generated.

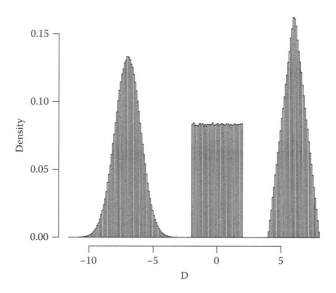

FIGURE 8.1
Illustration of normal, uniform, and triangular distributions.

You can generate random number sequences for probability distributions that are not already in R. Two methods we will cover are (1) the accept/reject method and (2) the transform method.

8.2.1 Accept/Reject Method

Suppose there is a probability density function (pdf) $p(x)$ defined between the finite values x_{min} and x_{max}. The maximum (peak) value of $p(x)$ is p_{max}. With this method, we take the following steps:

1. Generate two independent sets, x and y, of N uniformly distributed random numbers, where $x_{min} \leq x \leq x_{max}$ and $0 \leq y \leq p_{max}$.
2. Create the pairs $(x_1, y_1), (x_2, y_2),..., (x_i, y_i),..., (x_N, y_N)$.
3. Calculate $p(x_1), p(x_2),..., p(x_i),..., p(x_N)$.
4. If $y_i < p(x_i)$, retain point x_i in the set x.
5. If $y_i \geq p(x_i)$, discard point x_i from the set x.
6. Use the values remaining in x for the set of random numbers for $p(x)$.

We illustrate this procedure with the following R code:

```
> d <- function(x) {2*x}
> N <- 1e3
> x <- runif(n=N,min=0,max=1)
```

```
> y <- runif(n=N,min=0,max=max(d(x)))
> xind <- which(y<d(x))
> Nxind <- length(xind)
> xp <- numeric(Nxind)
> for (i in 1:Nxind) {xp[i] <- x[xind[i]]}
> hist(xp,prob=TRUE)
> curve(d(x),col="black",lwd=2,add=TRUE)
> quartz()
> plot(x,y)
> curve(d(x),col="black",lwd=2,add=TRUE)
```

Using this code, we generate a set of random numbers for the pdf $p(x) = 2x$, where $x_{min} = 0$ and $x_{max} = 1$. Figure 8.2 shows a plot of the random data pairs (x_i, y_i) as circles. The black line is the pdf $p(x) = 2x$, which is defined in the code using d <- function(x) {2*x}. The code identifies the data pairs that fall below the black line and the x_i values for these data pairs that are retained (with all others being rejected). Figure 8.3 shows the density plot for the retained x_i values (with the diagonal line in Figure 8.3 being the pdf). It is noted that Figures 8.2 and 8.3 were generated using $N = 10^3$ original data pairs. If we run the code again but with $N = 10^6$, we obtain Figure 8.4, which displays a better representation of the pdf.

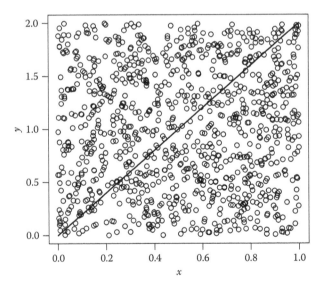

FIGURE 8.2
Plot of $N = 10^3$ random data pairs (circles). The black line is the pdf $p(x) = 2x$.

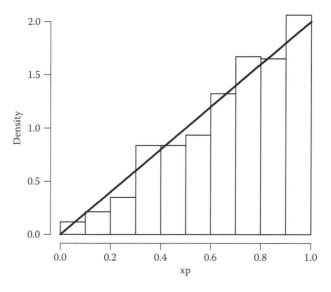

FIGURE 8.3
Density plot for the data points in Figure 8.4 that fall below the pdf. $N = 10^3$ here. The black diagonal line is the pdf $p(x) = 2x$.

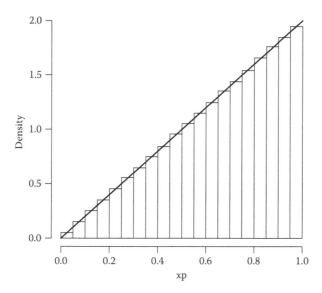

FIGURE 8.4
Density plot for $N = 10^6$. The black diagonal line is the pdf $p(x) = 2x$.

8.2.2 Inverse-cdf Method

With this method, we use cdfs to transform random numbers from one pdf to another. Let $p(x)$ be a pdf for x and $q(y)$ be a pdf for y. For random number x_i, we calculate the random number y_i by requiring that

$$\int_{y_{min}}^{y_i} q(y)\,dy = \int_{x_{min}}^{x_i} p(x)\,dx \tag{8.1}$$

holds. If $p(x)$ is a *uniform* distribution with $x_{min} = 0$ and $x_{max} = 1$, then Equation 8.1 can be written as

$$\int_{y_{min}}^{y_i} q(y)\,dy = x_i. \tag{8.2}$$

However, the left-hand side of Equation 8.2 is the cdf of y:

$$\text{cdf}_y(y_i) = \int_{y_{min}}^{y_i} q(y)\,dy \tag{8.3}$$

such that

$$y_i = \text{cdf}_y^{-1}(x_i) \tag{8.4}$$

holds. Equation 8.4 states that the random numbers x_i from a uniformly distributed set are mapped to y_i by applying the inverse cdf for y.

As an example, suppose $q(y) = 2y$ with $y_{min} = 0$ and $y_{max} = 1$. Equation 8.3 then yields $\text{cdf}_y(y_i) = y_i^2$. By Equation 8.4, we then have the mapping $y_i = x_i^{1/2}$. Here is the R code for this example:

```
> N <- 1e5                        # number of random numbers
> x <- runif(n=N,min=0,max=1)     # random numbers from a uniform pdf
> y <- sqrt(x)                    # random numbers for the pdf q = 2*y
> hist(y,prob=TRUE)
> yPlot <- seq(from=0,to=1,by=0.01)
> qPlot <- 2*yPlot
> lines(yPlot,qPlot,col="black",lwd=2)
```

Results of these calculations are shown in Figure 8.5 for an initial set of $N = 10^5$ uniformly distributed random numbers. The black diagonal line is the pdf $q(y) = 2y$. The density plot indicates that the mapping worked well for this example.

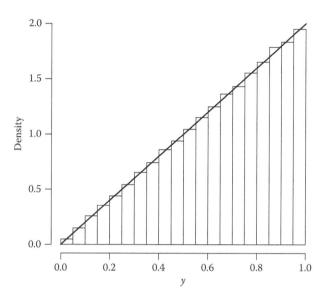

FIGURE 8.5
Density plot for $N = 10^5$. The black diagonal line is the pdf $q(y) = 2y$.

8.3 Random Sampling

We need to be able to randomly sample members of a data set. A way to do this with R is with the function `sample()`, which can sample both with and without replacement. If a data set is sampled with replacement, then any individual member of the set can show up in the sample more than once. If the data are sampled without replacement, then any member of the set can show up in the sample no more than once. Here is some R code to illustrate these concepts:

```
> x <- c(1,2,3,4,5,6,7,8,9,10)        # a data set
> nSample <- 5                        # number of elements in a sample
> sample(x,size=nSample,replace=TRUE)      # sample with replacement
[1] 2 3 6 6 2
> sample(x,size=nSample,replace=FALSE)     # sample without replacement
[1]   2   6   3 10   4
```

When we sampled with replacement (`replace=TRUE`), the numbers 2 and 6 each showed up twice, but without replacement (`replace=FALSE`), no number was repeated. The parameter `size` specifies the number of elements in the sample.

If you want to randomize a set of numbers, just use `sample()` without replacement and with `size` equal to the number of elements in the data set:

```
> x <- c(1,2,3,4,5,6,7,8,9,10)        # a data set
> nSample <- length(x)                # number of elements in a sample
> sample(x,size=nSample,replace=FALSE)   # sample without replacement
 [1]  4 10  3  1  2  7  6  8  9  5
```

8.4 Uncertainty of a Measured Variable

Suppose we have a sample of $N = 10$ data points and we want to calculate a 95% uncertainty interval for the mean of the population. The measured variable, x, is treated as a random variable, as follows:

$$x_n = \langle x \rangle + \alpha_n + \varepsilon_n + \beta, \tag{8.5}$$

where ε_n and β could be sums of random variables and the subscript n refers to data point n. The data for this example are from a normal population with $\mu = \langle x \rangle = 23$ and $\sigma^2 = V(\alpha) + V(\varepsilon) = 4$. The systematic errors have the variance $V(\beta) = b_x^2 = 0.5$. We will do this calculation using parametric bootstrapping, nonparametric bootstrapping, and Equation 5.37.

We use Equation 8.5 as a guide in setting up the MC algorithms for this example. The approach we will adopt involves the following steps:

1. Select the number of MC replicates, M, to be created.
2. For a given replicate, calculate N random values corresponding to $\langle x \rangle + \alpha_n + \varepsilon_n$. Also calculate a single random value for β and then apply Equation (8.5) to calculate x_n.
3. Calculate the statistic of interest, S, using the x_n values from step (2). Save this statistic's value in computer memory.
4. Repeat steps (2) and (3) M times, saving each S value.
5. Evaluate the $P\%$ coverage intervals for the values of the statistic that were saved.

The R code for this example is as follows:

```
> N <- 10                     # number of data points
> P <- 0.95                   # confidence level
> M <- 1e5                    # number of MC replicates
> trueMean <- 23
> trueSd <- 2
> options(digits=4)
> set.seed(100)
```

```
> x <- rnorm(n=N,mean=trueMean,sd=trueSd)
> bx <- sqrt(0.5)
>
> # parametric bootstrap
> xMean <- mean(x)
> xSd <- sd(x)
> S <- numeric(M)
> for (n in 1:M) {
+     beta <- rnorm(n=1,mean=0,sd=bx)
+     xSample <- rnorm(n=N,mean=xMean,sd=xSd)+beta
+     S[n] <- mean(xSample)
+ }
> quantile(S,probs=c((1-P)/2,(1+P)/2))
 2.5% 97.5%
21.42 24.52
> mean(S)
[1] 22.97
> hist(S,prob=TRUE,ylim=c(0,0.7),main="Parametric Bootstrap")
>
> # nonparametric bootstrap
> for (n in 1:M) {
+     beta <- rnorm(n=1,mean=0,sd=bx)
+     xSample <- sample(x,size=N,replace=TRUE)+beta
+     S[n] <- mean(xSample)
+ }
> quantile(S,probs=c((1-P)/2,(1+P)/2))
 2.5% 97.5%
21.42 24.52
> mean(S)
[1] 22.96
> quartz()
> hist(S,prob=TRUE,ylim=c(0,0.7),main="Nonparametric Bootstrap")
>
> # chapter 5 approach
> uxbar <- sqrt(xSd*xSd/N+bx^2)
> dof <- N-1                        # degrees of freedom
> k <- qt((P+1)/2,dof)              # coverage factor
> xMean-k*uxbar
[1] 21.17
> xMean
[1] 22.96
> xMean+k*uxbar
[1] 24.75
```

The density plots from the parametric and nonparametric bootstrap calculations in Figures 8.6 and 8.7, respectively, are very similar. The uncertainty intervals for the population mean are listed in Table 8.1.

Repeating the calculations with $N = 100$ data points yields the results in Table 8.2.

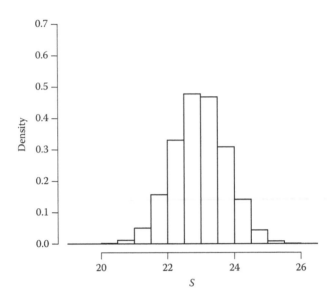

FIGURE 8.6
Density plot for a parametric bootstrap.

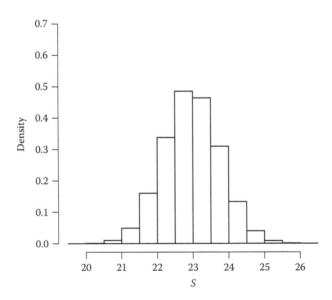

FIGURE 8.7
Density plot for a nonparametric bootstrap.

TABLE 8.1

Uncertainty Intervals for the Population Mean
for $N = 10$

Parametric	$\mu = 22.97 \pm 1.55$ (95%)
Nonparametric	$\mu = 22.96 \pm 1.54$ (95%)
Equation 5.37	$\mu = 22.96 \pm 1.79$ (95%)

TABLE 8.2

Uncertainty Intervals for the Population Mean
for $N = 100$

Parametric	$\mu = 23.01 \pm 1.45$ (95%)
Nonparametric	$\mu = 23.01 \pm 1.45$ (95%)
Equation 5.37	$\mu = 23.01 \pm 1.46$ (95%)

The results of all three approaches are similar, especially for $N = 100$. This is because we were calculating a confidence interval for the mean, and the distribution for the sampled means should be approximately normal per the central limit theorem, so Equation 5.37 would be expected to perform well in this situation.

8.5 Bootstrapping with Internal Functions in R

Bootstrapping can also be done using *internal* functions in R [3]. Parametric and nonparametric approaches are both available. To use these functions, you need to load the package `boot`. Here is the code for a nonparametric confidence interval for the mean of the data set in the previous example:

```
> library(boot)
> N <- 10                    # number of data points
> P <- 0.95                  # confidence level
> M <- 1e5                   # number of MC replicates
> trueMean <- 23
> trueSd <- 2
> options(digits=4)
> set.seed(100)
> x <- rnorm(n=N,mean=trueMean,sd=trueSd)
> S <- function(x,d) {
+     bx <- sqrt(0.5)
+     beta <- rnorm(n=1,mean=0,sd=bx)
```

```
+        mean(x[d])+beta
+        }
> bootStat <- boot(data=x,statistic=S,R=M)
> quartz()
> hist(bootStat$t[,1],prob=TRUE,ylim=c(0,0.7),main="R Internal Bootstrap Function")
> mean(bootStat$t[,1])
[1] 22.96
> boot.ci(bootStat,type="perc",conf=P)
BOOTSTRAP CONFIDENCE INTERVAL CALCULATIONS
Based on 100000 bootstrap replicates


CALL :
boot.ci(boot.out = bootStat, conf = P, type = "perc")


Intervals :
Level      Percentile
95%     (21.43, 24.51)
Calculations and Intervals on Original Scale
```

The statistic to be evaluated needs to actually be in a function (designated as S in the code). This function is referenced in the argument list of the function boot(), which calculates the replicates. The results from boot() are stored

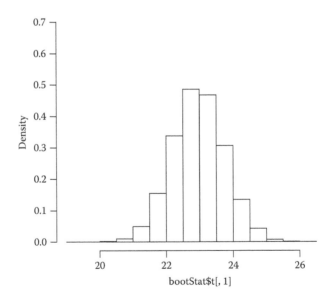

FIGURE 8.8
Density plot for a nonparametric bootstrap.

in the object `bootStat` and a coverage interval and a histogram are then generated. The coverage interval is evaluated using the internal function `boot.ci()`. By using the `type="perc"` parameter, we have instructed R to calculate the confidence level in the same way as the default `quantile()` function. The variable `bootStat$t[,1]` contains the mean values from the bootstrapping loop.

The confidence interval from the R internal bootstrap function is

$$\mu = 22.96 \pm 1.54 \, (95\%),$$

which is in reasonable agreement with our prior calculations for $N = 10$. The density plot for this bootstrap calculation (Figure 8.8) is also similar to the prior calculations (Figure 8.7).

8.6 Monte Carlo Convergence Criteria

Because an MC calculation uses random numbers, each replicate can be expected to be different than the others. This means that there will generally be fluctuations in the results no matter how many MC iterations are performed, so it is not always clear when a calculation has "converged." An approach [2] to decide when to stop MC calculations is to monitor the standard deviation of the statistic of interest as the calculations proceed and to stop when fluctuations in this standard deviation are small, e.g., in the range of 1%–5%. Another approach [1] is to specify M in advance of a calculation by using

$$M > \frac{10^4}{1-P}, \tag{8.6}$$

where P is the confidence level of interest. For example, for $P = 0.05$, Equation 8.6 yields $M > 2 \times 10^5$. The calculation times for large values of M can be long, which is why some of the calculations in this text use M values smaller than what is stipulated by Equation 8.6.

8.7 Uncertainty of a Result Calculated Using Experimental Data

The MC approach can be used to propagate uncertainties in a straightforward way even when there are systematic errors present. MC methods have several advantages [1] over the Taylor series approach discussed earlier for the propagation of uncertainties:

1. MC methods generate a pdf for the output variable, and this output pdf does not need to be normal. A normal pdf or a scaled and shifted t pdf [1] is implicitly assumed to be the case for the Taylor series method.
2. MC methods do not require specification of a coverage factor.
3. MC methods work well with nonlinear equations, where Taylor series approaches may not be as accurate.
4. MC methods can often be readily used for complicated situations in which it is very difficult to use a Taylor series approach.
5. MC methods work when the sensitivity coefficients are zero. Higher-order terms must be retained in the Taylor series expansion for this situation.

As an example, consider the situation in which we measure the lengths of two sides, x and y. Side x is measured 10 times using a set of calipers and side y is measured 10 times using a *different* set of calipers. The resulting data are listed in Table 8.3. We want to use this information to calculate the difference $r = y - x$ as well as its 95% confidence interval.

Our measurement systems are known to have the elemental systematic errors (at the 95% level) listed in Table 8.4. Both calipers were calibrated

TABLE 8.3

Caliper Measurement Data

x (mm)	9.93	9.91	10.14	10.19	10.00	10.01	10.02	10.03	10.11	10.01
y (mm)	19.87	20.07	20.10	19.98	20.21	20.07	19.88	20.08	20.01	19.86

TABLE 8.4

Systematic Errors of the Measurement System

	Errors for Calipers x and y (mm)
Linearity	0.01
Repeatability	0.005
Calibration	0.005

using the *same* standard, so this elemental error is *common* to both sets of measurements. The other elemental errors are *not* common.

For the linearity and repeatability errors, we generate different random numbers for x and y because they are assumed to be uncorrelated. For the calibration error, which is *common* to both x and y in this example, we generate a random number for x and use the *same* random number for y. We use normal distributions for all of the systematic elemental errors, but we could easily program in other distributions (e.g., uniform or triangular). The R code is as follows:

```
> x <- c(9.93,9.91,10.14,10.19,10.00,10.01,10.02,10.03,10.11,10.01)
> y <- c(19.87,20.07,20.10,19.98,20.21,20.07,19.88,20.08,20.01,19.86)
> Nx <- length(x)        # number of data points in x
> Ny <- length(y)        # number of data points in y
> P <- 0.95              # confidence level
> R <- 10000             # number of times to resample the data
> bLin <- 0.01/2         # linearity
> bRep <- 0.005/2        # repeatability
> bCal <- 0.005/2        # calibration
>
> boot.r <- numeric(R)   # vector for r values
>
> for (i in 1:R) {
+      boot.sample.x <- sample(x,size=Nx,replace=T)
+      boot.sample.y <- sample(y,size=Ny,replace=T)
+      beta1x <- rnorm(n=1,mean=0,sd=bLin)
+      beta2x <- rnorm(n=1,mean=0,sd=bRep)
+      beta3x <- rnorm(n=1,mean=0,sd=bCal)
+      beta1y <- rnorm(n=1,mean=0,sd=bLin)
+      beta2y <- rnorm(n=1,mean=0,sd=bRep)
+      beta3y <- beta3x
+      xs <- mean(boot.sample.x)+beta1x+beta2x+beta3x
+      ys <- mean(boot.sample.y)+beta1y+beta2y+beta3y
+      boot.r[i] <- ys-xs
+      }
>
> hist(boot.r)
> mean(boot.r)
[1] 9.9784
> quantile(boot.r, probs = c((1-P)/2,(1+P)/2))
     2.5%      97.5%
 9.891489 10.064263
```

The code yields

$$\bar{r} = 9.98 \pm 0.09 \ \text{mm} \left(95\% \right),$$

which is the same as calculated in Section 6.1 using the Taylor series approach for this same problem. The MC approach has several advantages:

It is usually simple to apply, we can use different distributions if needed, and we do not need to make any assumptions about the coverage factor k. Note that, instead of resampling, we could have employed a parametric bootstrap method.

We will now compare the MC and Taylor series approaches for propagating uncertainties into a highly nonlinear equation. This is done using an example from chemical kinetics, i.e., evaluation of the uncertainty in an Arrhenius rate constant K of the form

$$K = A\left(\frac{T}{T_{ref}}\right)^b e^{-E/RT},\tag{8.7}$$

where
   A is the pre-exponential factor
   b is a constant
   E is the activation energy
   T is the absolute temperature
   T_{ref} is a reference (absolute) temperature
   R is the universal gas constant

The variables A, b, and E are often determined from experiments and will thus have uncertainties associated with them. The universal gas constant R also has an uncertainty.

The Taylor series approach yields

$$\frac{u_K}{K} = \left[\left(\frac{u_A}{A}\right)^2 + \left(\frac{E}{RT}+b\right)^2\left(\frac{u_T}{T}\right)^2 + \ln^2\left(\frac{T}{T_{ref}}\right)(u_b)^2 + \left(\frac{E}{RT}\right)^2\left(\frac{u_E}{E}\right)^2 + \left(\frac{E}{RT}\right)^2\left(\frac{u_R}{R}\right)^2\right]^{1/2}\tag{8.8}$$

for the standard uncertainty in K, where we have assumed that covariances between A, b, T, and E are negligible and also that the standard uncertainties in these variables have been evaluated. We have also assumed that T_{ref} is defined without any uncertainty.

A gas-phase reaction that can occur in combustion systems is $O_2 \rightarrow O + O$. Parameter values listed for this reaction vary by source, and the NIST Chemical Kinetics Database [5] provides data in this regard. From this database, we can obtain the values $A = 1.01 \times 10^{-8}$, $b = -1.00$, $T_{ref} = 298$ K, and $E = 494$ kJ/mol that correspond to a particular experiment. We will assume that the relative standard uncertainties for all variables are 0.01

(i.e., 1%) except for *R*, which is 5.7×10^{-7}. The following code shows calculations for a temperature of 2000 K:

```
> # chemical kinetic rate constant uncertainty comparison
> N <- 1000000
> P <- 0.95
> Temp <- 2000
> Tref <- 298
> E <- 494
> A <- 1.01e-8
> b <- -1.00
> R <- 8.3144598e-3
> uTemp <- 0.01*Temp
> uE <- 0.01*E
> uA <- 0.01*A
> ub <- 0.01*abs(b)
> uR <- (5.7e-7)*R
>
> # Taylor series approach
> KBarTS <- A*((Temp/298)^b)*exp(-E/(R*Temp))
> uK <- KBarTS*sqrt((uA/A)^2+(E/R/Temp+b)^2*(uTemp/Temp)^2+log(Temp/
  Tref)^2*(ub)^2+(E/R/Temp)^2*(uE/E)^2+(E/R/Temp)^2*(uR/R)^2)
> t_value <- qt((P+1)/2,N-2)
> UKTS <- t_value*uK
> KLowerTS <- KBarTS-UKTS
> KUpperTS <- KBarTS+UKTS
>
> # Monte Carlo approach
> Temp <- rnorm(n=N,mean=Temp,sd=uTemp)
> A <- rnorm(n=N,mean=A,sd=uA)
> E <- rnorm(n=N,mean=E,sd=uE)
> b <- rnorm(n=N,mean=b,sd=ub)
> R <- rnorm(n=N,mean=R,sd=uR)
> K <- A*((Temp/298)^b)*exp(-E/(R*Temp))
> KLowerMC <- quantile(K,probs=c((1-P)/2,(1+P)/2))[[1]]
> KUpperMC <- quantile(K,probs=c((1-P)/2,(1+P)/2))[[2]]
> KBarMC <- mean(K)
> UKMC <- (KUpperMC-KLowerMC)/2
> d <- density(K)
> KMC_Most_Probable <- d$x[which(d$y==max(d$y))]
>
> # plot the results
> hist(K,prob=TRUE,n=100,xlim=c(0,3.5)*KMC_Most_Probable)
> d <- density(K)
> lines(d,col="black",lwd=2)
> abline(v=KBarTS,col="black",lwd=2)               # Taylor series
                                                     mean
> abline(v=KLowerTS,col="black",lwd=2,Hy="dashed")  # Taylor series
                                                     lower limit
> abline(v=KUpperTS,col="black",lwd=2,Hy="dashed")  # Taylor series
                                                     upper limit
> abline(v=KBarMC,col="gray",lwd=2)                 # Monte Carlo
                                                     mean
> abline(v=KLowerMC,col="gray",lwd=2,Hy="dashed")   # Monte Carlo
                                                     lower limit
```

```
> abline(v=KUpperMC,col="gray",lwd=2,lty="dashed")        # Monte Carlo
                                                             upper limit
> abline(v=KMC_Most_Probable,col="gray",lwd=2,lty="dotted")  # Monte Carlo
                                                             most probable
                                                             value

>
> # print the results
> KLowerTS
[1] 3.5707e-23
> KUpperTS
[1] 3.417166e-22
> KBarTS
[1] 1.887118e-22
> KLowerMC
[1] 8.287285e-23
> KUpperMC
[1] 4.199733e-22
> KBarMC
[1] 2.049222e-22
> KMC_Most_Probable
[1] 1.595121e-22
> UKTS/KBarTS
[1] 0.8107855
> UKMC/KBarMC
[1] 0.8225083
```

With the code, we calculate the 95% uncertainties for K and a density plot is also shown in Figure 8.9.

The black line at the bin tops in Figure 8.9 is a kernel density plot that shows the estimated pdf for the MC values that were calculated. The pdf is asymmetric such that the most probable K value, as marked by the dotted gray line, is

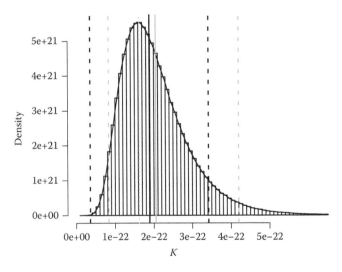

FIGURE 8.9
Density plot for the rate constant K.

smaller than the mean values from the Taylor series and MC calculations, i.e., the solid black and gray lines, respectively. The asymmetry in the pdf is also responsible for the differences in the upper and lower 95% uncertainty interval limits for the Taylor series and the MC calculations (the dashed black and gray lines, respectively). The Taylor series 95% interval for K is $(1.9 \pm 1.5) \times 10^{-22}$ and the corresponding 95% MC interval is $(2.5 \pm 1.7) \times 10^{-22}$. The MC approach is to be preferred for these types of calculations with highly nonlinear equations, which can produce pdfs that are not normal or scaled and shifted t distributions.

8.8 Uncertainty Bands for Linear Regression Curve Fits

MC methods offer advantages of simplicity and flexibility in developing confidence bands and for evaluating coefficient coverage intervals for linear regression curve fits. A disadvantage is that these methods can require significant computational resources, but complex problems can still be addressed with modern personal computers.

The approach we will adopt will involve the following steps:

1. For a given set of data (x_i, y_i), select the type of function we want to fit to the data, e.g., a straight line or a quadratic function.
2. Determine the coefficients a_0, a_1, ... via computer modeling. Save these values in computer memory.
3. Determine the predicted y_c values for the x values of interest. Save these y_c values in computer memory.
4. Resample the (x_i, y_i) data pairs with replacement. If needed, for each resampled data pair, add in random values to account for systematic errors.
5. Repeat steps (2)–(4) a large number of times, e.g., at least 1000 times.
6. Evaluate the $P\%$ coverage intervals for the y_c values that were saved. These coverage intervals are the coverage band limits for each x value.
7. Evaluate the $P\%$ coverage intervals for the coefficient values a_0, a_1, ... that were saved.

Note that, instead of resampling in step (4), we could instead use a parametric bootstrap, but to do so we would need to specify the pdf as well as statistics such as a mean and a standard deviation.

We illustrate these steps with the following R code:

```
> x <- seq(from=1,to=10,length.out=20)    # x data
> N <- length(x)                          # number of data pairs
> yTrue <- 1 + x                          # true y data
> y <- yTrue + 1.0*rnorm(N)               # noisy y data
> P <- 0.95                               # confidence level
```

```
> plot(x,y,type="n",xlim=c(0,12),ylim=c(0,12))  # create a plot window
> i <- seq(from=1,to=N,by=1)                # data pair identifiers
> R <- 100000                               # number of replicates
> ymc <- matrix(nrow=R,ncol=N)       # matrix to hold y(x) fit data
> a0 <- numeric(R)                   # vector to hold a0 data
> a1 <- numeric(R)                   # vector to hold a1 data
> for (j in 1:R){
      # random sample of data identifiers
      jSample <- sample(i,size=length(x),replace=TRUE)
      xSample <- x[jSample]        # random sample of data points
      ySample <- y[jSample]        # random sample of data points
      fitj <- lm(ySample~xSample)  # generate a fit
      a0[j] <- fitj$coefficients[[1]]      # extract a0
      a1[j] <- fitj$coefficients[[2]]      # extract a1
      # save y(x) for this fit
      for (k in 1:N) {ymc[j,k] <- a0[j]+a1[j]*x[k]}
      # prediction object
      fitjPredict <- predict(fitj,interval = "confidence",level=P)
      }
>
> # plot the original data points
> points(x,y)
>
> # plot the true data
> lines(x,yTrue,col="black")
>
> # save the upper and lower Monte Carlo data
> lwr <- numeric(N)
> upr <- numeric(N)
> for (n in 1:N) {
      ans <- quantile(ymc[,n], probs=c((1-P)/2,(1+P)/2))
      lwr[n] <- ans[[1]]
      upr[n] <- ans[[2]]
 }
>
> # plot the upper and lower Monte Carlo data
> lines(x,lwr,col="gray",lw=2)
> lines(x,upr,col="gray",lw=2)
>
> # plot a line fit to the original data
> fit1 <- lm(y~x)
> fit1Predict <- predict(fit1, interval = "confidence",level=P)
> lines(x,fit1Predict[,"fit"])
>
> # plot the confidence limits for the original data
> lines(x, fit1Predict[, "lwr"], lty = "dotted")
> lines(x, fit1Predict[, "upr"], lty = "dotted")
>
> # print the mean Monte Carlo a0 and a1 values
> mean(a0)
[1] 0.8631118
> mean(a1)
[1] 0.9943353
>
> # print the a0 and a1 values from the line fit to the original data
> fit1$coefficients
```

```
(Intercept)             x
  0.8705913    0.9943024
>
> # print the Monte Carlo a0 and a1 confidence intervals
> quantile(a0,probs=c((1-P)/2,(1+P)/2))
      2.5%        97.5%
0.1707875 1.6284774
> quantile(a1,probs=c((1-P)/2,(1+P)/2))
      2.5%        97.5%
0.8790259 1.0963466
>
> # print the a0 and a1 P% confidence intervals using R
> confint(fit1,level=P)
                   2.5 %      97.5 %
(Intercept) 0.2056191 1.535564
x               0.8860165 1.102588
```

The results are shown in Figure 8.10. In this code, we add normal noise to a straight line. A 95% confidence band is fit to the data using the `predict()` function (the dotted lines) and the 95% confidence band from the MC approach is shown as the gray lines. We also calculate values for a_0 and a_1 as well as for their confidence intervals. The agreement between the two approaches is quite good.

In this example, we did not consider systematic errors. This was so that we could compare the `predict()` and MC results, but it would have been straightforward to include systematic errors by adding the appropriate random numbers to each resampled data set.

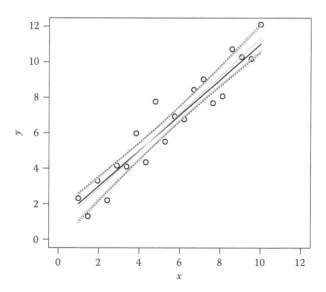

FIGURE 8.10
Comparison of MC (gray lines) and `predict()` (dotted lines) results for curve-fit confidence bands.

8.9 Uncertainty Bands for a Curve Fit with Kernel Smoothing

We now consider the use of bootstrapping for generating uncertainty bands for a curve fit to noisy data. The curve is fit with a kernel smoothing technique, i.e., `locpol()`, which makes it difficult to apply a Taylor series approach. We will generate uncertainty bands for the function $y = \sin(x)$ and its first derivative $dy/dx = \cos(x)$. The function y has normal noise added to it.

Our approach is to resample the data points with replacement and calculate a `locpol()` fit for each set of resampled data. The fitted curve-fit values, denoted here as Y, are saved at specific values of x and then the `quantile()` function is used to calculate the uncertainty band limits for Y. To generate the uncertainty band, these limits are plotted above and below the mean Y values from the fits. The R code is as follows:

```
> library(locpol)
> bwFit <- 0.3                              # bandwidth
> R <- 10000                                # number of replicates
> P <- 0.95                                 # confidence level
> x <- seq(from=0,to=10,length.out=301)     # x data
> yTrue <- sin(x)                           # true y data
> dyTruedx <- cos(x)                        # true dydx data
> set.seed(100)                             # random number generator seed
> y <- yTrue + 0.1*rnorm(n=length(yTrue))      # noisy y data
> plot(x,y,type='n')                        # open a plot window
> N <- length(x)                            # number of data pairs
> I <- 1:N                                  # integer sequence for
                                              sampling indices
> xSample <- numeric(N)                     # vector to hold sampled x
                                              values
> ySample <- numeric(N)                     # vector to hold sampled y
                                              values
> Y <- matrix(nrow=R,ncol=N)                # matrix to hold Y for each fit
> dYdx <- matrix(nrow=R,ncol=N)             # matrix to hold dYdx for each
                                              fit
>
> # bootstrapping loop
> for (j in 1:R) {
+     jSample <- sort(sample(I,size=N,replace=TRUE))
+     xSample <- x[jSample]
+     ySample <- y[jSample]
+     data <- data.frame(xSample,ySample)
+     # fit the data - use xeval=x to have the right order for the
        fitted data
+     fitj <- locpol(ySample~xSample,data=data,deg=2,xeval=x,kernel=gaussK,
        bw=bwFit)
+     Xj <- fitj$xeval
+     Yj <- fitj$lpFit$ySample             # save the current Y values
+     dYjdx <- fitj$lpFit$ySample1         # save the current dYdx values
+     # lines(Xj,Yj)                        # plot the current fit (this
                                              can be slow)
```

```
+      for (k in 1:N) {Y[j,k] <- Yj[k]}              # save Y for this fit
+      for (k in 1:N) {dYdx[j,k] <- dYjdx[k]}        # save dYdx for this fit
+ }
>
> # evaluate the Y confidence band data
> YTop <- numeric(N)                     # vector for upper conf band limit
> YBot <- numeric(N)                     # vector for lower conf band limit
> YMean <- numeric(N)                    # vector for mean of the conf bands
> for (i in 1:N) {YBot[i] <- quantile(Y[,i],probs=c((1-P)/2,(1+P)/2),na.
  rm=TRUE)[[1]]}
> for (i in 1:N) {YTop[i] <- quantile(Y[,i],probs=c((1-P)/2,(1+P)/2),na.
  rm=TRUE)[[2]]}
> for (i in 1:N) {YMean[i] <- mean(Y[,i])}
>
> # evaluate the dYdx confidence band data
> dYTopdx <- numeric(N)                  # vector for upper conf band limit
> dYBotdx <- numeric(N)                  # vector for lower conf band limit
> dYMeandx <- numeric(N)                 # vector for mean of the conf bands
> for (i in 1:N) {dYBotdx[i] <- quantile(dYdx[,i],probs=c((1-P)/2,
  (1+P)/2),na.rm=TRUE)[[1]]}
> for (i in 1:N) {dYTopdx[i] <- quantile(dYdx[,i],probs=c((1-P)/2,
  (1+P)/2),na.rm=TRUE)[[2]]}
> for (i in 1:N) {dYMeandx[i] <- mean(dYdx[,i])}
>
> # plot the Y results
> points(x,y,pch=".",cex=4)             # plot the original data points
> lines(x,yTrue,col="black",lwd=2)      # plot the true data
> lines(x,YMean,col="black",lwd=2,Hy="dashed")        # plot the mean
                                                        fit
> lines(x,YTop,col="gray",lwd=2,Hy="dashed")          # plot the upper
                                                        conf band limit
> lines(x,YBot,col="gray",lwd=2,Hy="dashed")          # plot the lower
                                                        conf band limit
>
> # plot the dYdx results
> quartz()
> plot(x,dyTruedx,col="black",lwd=2,type='l',ylab="dy/dx") # plot the true
                                                            data
> lines(x,dYMeandx,col="black",lwd=2,Hy="dashed")    # plot the mean
                                                        fit
> lines(x,dYTopdx,col="gray",lwd=2,Hy="dashed")      # plot the upper
                                                        conf band limit
> lines(x,dYBotdx,col="gray",lwd=2,Hy="dashed")      # plot the lower
                                                        conf band limit
```

The results of the calculations are shown in Figures 8.11 and 8.12 for 10,000 replications of the data. The first plot is y as a function of x. The points are the original noisy data and the solid black and dashed black lines are the original data and the mean fitted line. These two lines overlap and it is difficult to distinguish between them on the plot. The upper and lower uncertainty band limits are the two dashed gray lines. The uncertainty band is generally thin except near the peaks and valleys of the data, which is where the noise appears to be more prevalent.

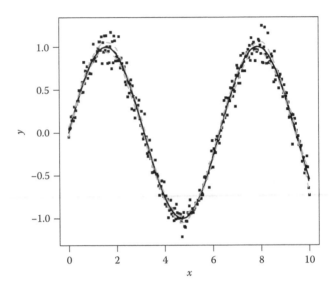

FIGURE 8.11
Uncertainty band (dashed gray lines) for a `locpol()` fit to noisy data.

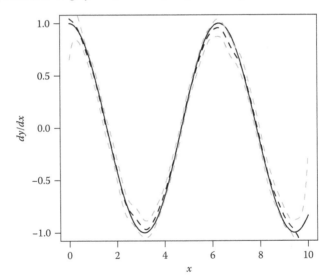

FIGURE 8.12
Uncertainty band (dashed gray lines) for the first derivative of noisy data obtained using `locpol()`.

Figure 8.12 shows dy/dx as a function of x. There is a good agreement between the mean fitted values and the noise-free function except near the endpoints. The uncertainty band limits are widest near the endpoints and also where the noise is apparently at its highest values, i.e., near the peaks and valleys. The uncertainty band is on the average wider for dy/dx than for y, which is reasonable given that differentiation generally amplifies noise.

Problems

P8.1 Two variables, x and y, were measured in an experiment. These variables have the nominal values $x = 10$ and $y = -20$. Using Taylor series and MC approaches, estimate the 95% uncertainty in the variable z for each of the following data reduction equations if the standard uncertainties are $u_x = 0.02$ and $u_y = 0.05$:

a. $z = \ln[(1+x^2)y^2]$

b. $z = x^2 - y^2 - 20x - 40y - 290$

P8.2 Temperature measurements in degree Celsius were made with a sensor that is known to have a systematic error of 0.005 K (95%), yielding the following results:

0.23	0.35	0.16	0.19	0.26	0.25	0.23	0.18	0.26	0.29
0.32	0.24	0.38	0.25	0.27	0.24	0.21	0.21	0.29	0.26

a. Using a Monte Carlo approach, determine the combined standard uncertainty for the mean value of the temperature.

b. Using a Monte Carlo approach, determine the combined standard uncertainty for the temperature.

P8.3 The following data set was obtained using a measurement system that has two elemental systematic errors that are uncorrelated:

28.75	30.37	28.33	33.19	30.66	28.36	30.98	31.48	31.15	29.39
33.02	30.78	28.76	25.57	32.25	29.91	29.97	31.89	31.64	31.19

One of these errors has a uniform pdf with $\beta_{min} = -2$ and $\beta_{max} = 2$. The pdf for the other error is normal with a mean of zero and a standard deviation of 0.1. Calculate the combined standard uncertainty for the mean value of the data.

P8.4 A pdf $p(x)$ is given by the expression

$$p(x) = Ax(1-x), \quad 0 \leq x < 1.$$

Generate a set of at least 10^5 random numbers corresponding to this pdf using the accept/reject approach.

a. Calculate the mean.

b. Calculate the standard deviation.

c. Calculate the standard deviation of the mean for $N = 10, 10^2, 10^3$, and 10^4, where N is the number of values in a random sample.

Provide a histogram for your random numbers. (Use `prob=TRUE` and also plot $p(x)$ for comparison.) Note that A is a constant you need to specify.

P8.5 A pdf $p(x)$ is given by the expression

$$p(x) = A\sin(\pi x), \quad 0 \le x < 1.$$

Generate a set of at least 10^5 random numbers corresponding to this pdf using the inverse-cdf approach.

a. Calculate the mean.

b. Calculate the standard deviation.

c. Calculate the standard deviation of the mean for $N = 10, 10^2, 10^3$, and 10^4, where N is the number of values in a random sample.

Provide a histogram for your random numbers. (Use `prob=TRUE` and also plot $p(x)$ for comparison.) Note that A is a constant you need to specify.

P8.6 Use the following R code to generate two sets of data, x and y:

```
x <- seq(from=0,to=4,length.out=21)    # x data
yTrue <- 2+x^2                          # true y data
N <- length(x)                          # number of data pairs
set.seed(100)
y <- yTrue+rnorm(n=N,mean=0,sd=1)       # noisy y data
```

Then generate the best-fit line for the expression $y_c(x) = a_0 + x^{a_1}$ using `nls()`. Using a nonparametric bootstrapping approach with `nls()`, determine the 95% uncertainty band for this curve fit.

P8.7 Repeat P8.6, but use a local polynomial approach.

P8.8 For the results show in Figure 8.9, determine which variable (A, b, E, T, T_{ref}, R) contribute most strongly to the uncertainty in K. Do this for both the Taylor series and MC approaches.

P8.9 Solve P7.9 (the baseball problem) using a Monte Carlo approach.

P8.10 Experiments to measure the drag on a sphere of diameter D yield the following data for drag force F_D as a function of the fluid velocity V:

V (m/s)	0.01	0.02	0.03	0.04	0.05	0.06	0.07
F_D (N)		0.0026	0.0052	0.0079	0.011	0.013	0.015	0.018

Assume that the sphere is in a fluid with density ρ and use the following nominal values and relevant uncertainties:

V: see values in the given table, $b = 0.0005$ m/s; F_D: see values in the given table, $b = 0.00007$ N;

D: 0.01 m, b = 0.0001 m; ρ: 1260 kg/m^3, s = 1.260 kg/m^3; μ: 1.4 kg/ (m s); s = 0.014 kg/(m s).

Plot the drag coefficient $C_D = 4F_D/(\pi\rho V^2 D^2)$ as a function of the inverse of the Reynolds number Re $= \rho VD/\mu$, where μ is the dynamic viscosity. Also use lm() to fit an equation of the form $C_D = a_0 + a_1/\mathrm{Re}$ to the data. Plot your curve fit ($C_D = a_0 + a_1/\mathrm{Re}$) as well as its 95% uncertainty band.

References

1. JCGM 101:2008, Evaluation of measurement data: Supplement 1 to the "Guide to the expression of uncertainty in measurement"—Propagation of distributions using a Monte Carlo method, International Bureau of Weights and Measures (BIPM), Sérres, France, 2008.
2. H. W. Coleman and W. G. Steele, *Experimentation, Validation, and Uncertainty Analysis for Engineers*, 3rd edn., Wiley, Hoboken, NJ, 2009.
3. M. J. Crawley, *The R Book*, 2nd edn., Wiley, West Sussex, U.K., 2013.
4. M. R. Chernick and R. A. LaBudde, *An Introduction to Bootstrap Methods with Applications to R*, Wiley, Hoboken, NJ, 2011.
5. NIST, NIST chemical kinetics database, http://kinetics.nist.gov/kinetics/ KineticsSearchForm.jsp, 2013 (accessed December 31, 2016).

9

The Bayesian Approach

We now consider the Bayesian approach to estimating uncertainty. The literature on Bayesian theory is vast, but References [1–6] can be recommended as useful resources. The reader can consult these references for more information than what is presented here.

Understanding the Bayesian approach requires that we consider the concept of probability in more detail than before. We begin by defining some terms.

Experiment: An observation or measurement.

Trial: A single performance of an experiment.

Outcome or sample point: Result of a trial.

Elementary event (X_i): A possible outcome that is exclusive; i.e., the occurrence of a particular elementary event implies that none of the other elementary events occur.

Sample space (Ω): The set of all possible elementary events corresponding to an experiment.

Event: A subset of Ω. If an outcome is in subset A, then event A has occurred.

As an example, suppose we flip a coin twice. The coin can come up either heads (H) or tails (T) as an outcome of either flip. The sample space is Ω = {HH,HT,TH,TT}, where HH, HT, TH, and TT are all elementary events. The first letter of an elementary event signifies the outcome of the first flip and the second letter the outcome of the second flip; e.g., for HT, the first coin flip yields H and the second yields T. If we define the event A = {HH,HT}, then A has occurred if the coin flips yield HH or HT. If we know all of the possible outcomes, then the probability of an event A can be defined as $P(A)=N_A/N$, where N_A is the number of elementary events in A and N is the total number of possible elementary events. For our earlier example, $N_A=2$ and $N = 4$ such that $P(A) = 2/4 = 1/2$. We have assumed here that the elementary elements are equally probable, which is sometimes questionable.

Interestingly, there is no single interpretation of probability that is widely accepted. There are two main interpretations of probability: frequentist and

Bayesian. With the frequentist interpretation, the probability P of an event A is defined using a limiting process; i.e.,

$$\text{Frequentist}: P(A) = \lim_{n \to \infty} \frac{\text{number of times } A \text{ occurs}}{n},$$

where the number of experiments performed, n, goes to infinity. In practical experiments, it is not possible to observe something an infinite number of times, which is something that Bayesians object to.

The Bayesian interpretation is that an observer's estimate of the probability that an event A will occur is based upon all of the information the observer has, including the number of times A occurred in the past; i.e.,

Bayesian: $P(A) \propto$ number of times A already occurred + other information.

The Bayesian approach can be somewhat subjective, which is a reason why frequentists criticize it.

Regardless of which interpretation of probability you subscribe to, you do have the mathematical theory of probability at your disposal. At a basic level, Kolmogorov provided axioms [1] for the probabilities $P(X_i)$ of elementary events, as shown in the following equations:

$$P(X_i) \geq 0, \tag{9.1}$$

$$\sum_{\Omega} P(X_i) = 1, \tag{9.2}$$

and

$$P(X_i \text{ or } X_j) = P(X_i) + P(X_j). \tag{9.3}$$

Equation 9.1 states that all probabilities are nonnegative, Equation 9.2 states that the sum of the probabilities of all elementary events in Ω is unity, and Equation 9.3 states that the probability of the occurrence of either X_i or X_j from a given trial is the sum of their individual probabilities. The Kolmogorov axioms can be used to deduce other properties and expressions. For example, if we consider the probability of achieving X_i and then X_j in two different trials, and the occurrence of X_j is independent of whether X_i occurred first, then we can write

$$P(X_i \text{ then } X_j) = P(X_i)P(X_j). \tag{9.4}$$

Elementary events that satisfy Equation 9.4 are said to be *independent*.

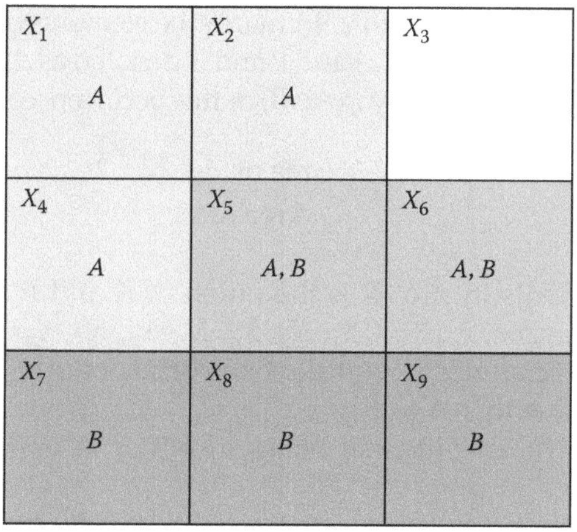

FIGURE 9.1
Venn diagram.

Venn diagrams provide a convenient way to visualize how probabilities of events in a sample space are related. For example, consider a sample space consisting of nine elementary events, i.e., $\Omega = \{X_1, X_2, X_3, X_4, X_5, X_6, X_7, X_8, X_9\}$, as illustrated in Figure 9.1. The large rectangle in Figure 9.1 contains all possible elementary events; i.e., it is Ω. The elementary events are represented by each of the smaller rectangles, and each elementary event is identified in the upper left-hand corner of each small rectangle. The probabilities of the elementary events are all the same; i.e., $P(X_i) = 1/9$. The rectangles that are light gray correspond only to event A, i.e., $\{X_1, X_2, X_4, X_5, X_6\}$, the dark gray rectangles correspond only to event B, i.e., $\{X_5, X_6, X_7, X_8, X_9\}$, and the medium gray rectangles correspond to both A and B, i.e., $\{X_5, X_6\}$. The area of region A is proportional to the size of the set of outcomes in event A, and the area of the large rectangle is proportional to the size of the set comprising the sample space Ω. It then follows that the probability $P(A)$ of the occurrence of event A is given by

$$P(A) = \frac{\text{Area of } A}{\text{Area of } \Omega} = \frac{5}{9}.$$ (9.5)

It similarly follows that

$$P(B) = \frac{\text{Area of } B}{\text{Area of } \Omega} = \frac{5}{9}$$ (9.6)

holds.

The medium gray region in Figure 9.1 marks the elements of Ω that are in both A and B. This is the intersection of A and B and is denoted as $A \cap B$. For Figure 9.1, $A \cap B = \{X_5, X_6\}$, and the probability $P(A \cap B)$ of this occurrence is given by

$$P(A \cap B) = \frac{\text{Area of } A \cap B}{\text{Area of } \Omega} = \frac{2}{9}.$$ (9.7)

The set of all elements in A or B is the union of A and B and is denoted as $A \cup B$. For this example, $A \cup B = \{X_1, X_2, X_4, X_5, X_6, X_7, X_8, X_9\}$. In Figure 9.1, $A \cup B$ is all small rectangles except the one corresponding to X_3, which was not defined to be in either A or B.

The probability of an outcome being in either A or B, i.e., $P(A \cup B)$, is given by

$$P(A \cup B) = \frac{\text{Area of } A + \text{Area of } A - \text{Area of } A \cap B}{\text{Area of } \Omega}$$

$$= P(A) + P(B) - P(A \cap B) = \frac{8}{9}.$$ (9.8)

The quantity $P(A \cap B)$ is subtracted in Equation 9.8 so that the elements in $A \cap B$ are not counted twice. In the case where the sets A and B are disjoint, i.e., they have no common elements, the following equation applies:

$$\text{Disjoint sets}: P(A \cup B) = P(A) + P(B).$$ (9.9)

The conditional probability $P(B|A)$ is the probability of the occurrence of event B on the condition that event A has occurred. This means that, if we have an elementary event that is in A, then $P(B|A)$ is the probability that this same elementary event is also in B. Sometimes, we state this as the probability of B given A where the vertical bar in $P(B|A)$ represents the word "given." From Figure 9.1, $P(B|A)$ is given by

$$P(B|A) = \frac{\text{Area of } A \cap B}{\text{Area of } A} = \frac{P(A \cap B)}{P(A)} = \frac{2/9}{5/9} = \frac{2}{5}.$$ (9.10)

The conditional probability $P(A|B)$ is given by

$$P(A|B) = \frac{\text{Area of } A \cap B}{\text{Area of } B} = \frac{P(A \cap B)}{P(B)} = \frac{2/9}{5/9} = \frac{2}{5}.$$ (9.11)

Equations 9.10 and 9.11 can be solved for $P(A \cap B)$ to yield

$$P(A \cap B) = P(B|A)P(A) = \frac{2}{9}$$ (9.12)

and

$$P(A \cap B) = P(A|B)P(B) = \frac{2}{9}. \tag{9.13}$$

By equating Equations 9.12 and 9.13, we obtain

$$P(A|B) = \frac{P(B|A)P(A)}{P(B)} = \frac{(2/5)(5/9)}{5/9} = \frac{2}{5}, \tag{9.14}$$

which is a statement of *Bayes' theorem*. Equation 9.14 can be interpreted as providing information about the probability of the occurrence of A given B by using prior information about the probability of A, the probability of B, and the probability of the occurrence of B given A. We would use Equation 9.14, e.g., if we did not already know $P(A|B)$.

Events A and B are defined to be independent if the following applies:

$$\text{Independent events: } P(A \cap B) = P(A)P(B). \tag{9.15}$$

Inserting Equation 9.15 into Equations 9.12 and 9.13 yields

$$\text{Independent events: } P(A|B) = P(A) \tag{9.16}$$

and

$$\text{Independent events: } P(B|A) = P(B). \tag{9.17}$$

Equations 9.16 and 9.17 state that the occurrence of one of the independent events (A or B) provides no new information on the probability of occurrence of the other event. In regard to Figure 9.1, events A and B are not independent.

If events A and B are disjoint, then the following hold:

$$\text{Disjoint events: } P(A|B) = 0 \tag{9.18}$$

and

$$\text{Disjoint events: } P(B|A) = 0. \tag{9.19}$$

In this case, the occurrence of one of the events (A or B) gives us complete information about the occurrence of the other event, i.e., that it certainly did not occur.

As an example of the use of conditional probability, consider a situation where pressure (p) and temperature (T) measurements of a reactor provide the following information:

A: The probability that a temperature measurement exceeds 373 K is 25%.

B: The probability that a pressure measurement exceeds 2 atm is 10%.

C: The probability that if $T > 373$ K, then $p > 2$ atm is 30%.

What is the probability that if $p > 2$ atm, then $T > 373$ K? To answer this question, we note that $P(A) = 0.25$, $P(B) = 0.10$, and $P(B\,|\,A) = 0.30$. We then solve for $P(A\,|\,B) = 0.75$, i.e., the probability that if $p > 2$ atm, then $T > 373$ K is 75%.

9.1 Bayes' Theorem for Probability Density Functions

Consider a multivariate probability density function (pdf) (see the appendix) that is a function of L random variables, i.e., $p(x_1, x_2, \ldots, x_L)$. We define the vectors $\vec{X}$ and $\vec{\Theta}$ in

$$\vec{X} = \left(x_1, x_2, \ldots, x_j \right) \tag{9.20}$$

and

$$\vec{\Theta} = \left(x_{j+1}, x_{j+2}, \ldots, x_L \right) \tag{9.21}$$

such that our pdf can be written as

$$p\left(x_1, x_2, \ldots, x_L \right) = p\left(\vec{X}, \vec{\Theta} \right). \tag{9.22}$$

We define the functions $g\left(\vec{X} \right)$ and $h\left(\vec{\Theta} \right)$ by integrating Equation 9.22 over $\vec{X}$ and $\vec{\Theta}$ separately:

$$g\left(\vec{X} \right) = \int_{\vec{\Theta}} p\left(\vec{X}, \vec{\Theta} \right) d\vec{\Theta} \tag{9.23}$$

and

$$h\left(\vec{\Theta} \right) = \int_{\vec{X}} p\left(\vec{X}, \vec{\Theta} \right) d\vec{X}. \tag{9.24}$$

We now hold all of the $\vec{\Theta}$ values constant with these values denoted as $\vec{\Theta}_0$ such that $p(\vec{X},\vec{\Theta}) = f(\vec{X},\vec{\Theta}_0)$. We use the notation "$f$" here because if we hold $\vec{\Theta}$ constant, then $p(\vec{X},\vec{\Theta})$ is no longer a pdf because it is not normalized if we integrate only over $\vec{X}$. If we normalize $f(\vec{X},\vec{\Theta}_0)$ to produce a pdf, we obtain

$$p(\vec{X}|\vec{\Theta}_0) = \frac{f(\vec{X},\vec{\Theta}_0)}{h(\vec{\Theta}_0)}. \tag{9.25}$$

The left-hand side of Equation 9.25 is the conditional pdf for $\vec{X}$ given $\vec{\Theta}_0$. However, the $\vec{\Theta}_0$ values are arbitrary so we can drop the subscript 0 in Equation 9.25, yielding

$$p(\vec{X}|\vec{\Theta}) = \frac{p(\vec{X},\vec{\Theta})}{h(\vec{\Theta})}. \tag{9.26}$$

If we follow the same procedure for $\vec{\Theta}$, holding all of the $\vec{X}$ values constant, we can derive

$$p(\vec{\Theta}|\vec{X}) = \frac{p(\vec{X},\vec{\Theta})}{g(\vec{X})}. \tag{9.27}$$

The term $p(\vec{X},\vec{\Theta})$ can be isolated in Equations 9.26 and 9.27 to yield

$$p(\vec{X},\vec{\Theta}) = p(\vec{X}|\vec{\Theta})h(\vec{\Theta}) = p(\vec{\Theta}|\vec{X})g(\vec{X}). \tag{9.28}$$

Rearranging Equation 9.28 yields

$$p(\vec{\Theta}|\vec{X}) = \frac{p(\vec{X}|\vec{\Theta})h(\vec{\Theta})}{g(\vec{X})}, \tag{9.29}$$

which is Bayes' theorem for pdfs. The use of Equations 9.23 and 9.28 in Equation 9.29 yields another form of Bayes' theorem for pdfs:

$$p(\vec{\Theta}|\vec{X}) = \frac{p(\vec{X}|\vec{\Theta})h(\vec{\Theta})}{\int_{\vec{\Theta}} p(\vec{X}|\vec{\Theta})h(\vec{\Theta})d\vec{\Theta}}. \tag{9.30}$$

We name the terms in Equation 9.30 as follows:

$p\left(\vec{\Theta}\mid\vec{X}\right)$ is the posterior pdf for $\vec{\Theta}$ (or simply the "posterior").

$p\left(\vec{X}\mid\vec{\Theta}\right)$ is the likelihood function for $\vec{\Theta}$ (or simply the "likelihood").

$h\left(\vec{\Theta}\right)$ is the prior function for $\vec{\Theta}$ (or simply the "prior").

$\int_{\vec{\Theta}} p\left(\vec{X}\mid\vec{\Theta}\right)h\left(\vec{\Theta}\right)d\vec{\Theta}$ is a normalization factor.

Equation 9.30 can be expressed as

$$p\left(\vec{\Theta}\mid\vec{X}\right)=c\,p\left(\vec{X}\mid\vec{\Theta}\right)h\left(\vec{\Theta}\right),\tag{9.31}$$

where the normalization constant c in Equation 9.31 is given by

$$c=\frac{1}{\int_{\vec{\Theta}} p\left(\vec{X}\mid\vec{\Theta}\right)h\left(\vec{\Theta}\right)d\vec{\Theta}}=\frac{1}{\int_{\vec{\Theta}} p\left(\vec{X},\vec{\Theta}\right)d\vec{\Theta}}.\tag{9.32}$$

A common application of Equation 9.31 is for the case where we want to estimate $\vec{\Theta}$ values when the values of the components of $\vec{X}$ (i.e., x_i) are independent. In this case, the likelihood function for $\vec{\Theta}$ is the product of the individual pdfs for each x_i:

$$p\left(\vec{X}\mid\vec{\Theta}\right)=\prod_i p_i\left(x_i,\vec{\Theta}\right).\tag{9.33}$$

In Equation 9.33, $p_i\left(x_i,\vec{\Theta}\right)$ is the pdf for variable x_i.

Priors are arguably the most discussed aspects of Bayesian analysis. A prior is supposed to reflect our knowledge about $\vec{\Theta}$ before the newest data $\vec{X}$ are accounted for in Equation 9.31. If information about $\vec{\Theta}$ already exists, then we would need to incorporate this into $h\left(\vec{\Theta}\right)$ [3,7]. If, however, we have no reliable prior information about $\vec{\Theta}$, it is recommended that a reference prior be used that reflects our lack of knowledge; i.e., it should be uninformative. Jeffreys [8] suggested that an uninformative prior should be unity for a "location" parameter such as μ or the inverse of a "scale" parameter such as σ; i.e., use $1/\sigma$ if you want to evaluate the pdf for σ. If you want to evaluate both μ and σ, the reference prior $h(\mu,\sigma)=1/\sigma$ can be used for this as well. More complex cases are discussed in Reference [9].

It is often the case that a reference prior is improper in that it cannot be normalized (integrated to unity). This is generally not a practical difficulty, however, as the posterior is usually proper; i.e., it integrates to unity. We also note that as the number of data points increases, the influence of the specific functional form of $h\left(\vec{\Theta}\right)$ typically becomes small in the sense that the posterior

$p\left(\vec{\Theta}\middle|\vec{X}\right)$ becomes independent of the prior. It is worthwhile when solving a problem to actually use different priors to see how they influence the results.

9.2 Bayesian Estimation of the Mean and Standard Deviation of a Normal Population

Here, we will use a Bayesian approach to evaluate pdfs for the mean and standard deviation of a normal population for the situation in which the data have a systematic error β with standard deviation b and a mean of zero. To proceed, we will let $\vec{X} = \{x_1, x_2, \ldots, x_N\}$, where $x_1, x_2, \ldots$ are N data values. In addition, $\vec{\Theta} = \{\mu, \sigma, \beta\}$.

We assume that the components of $\vec{X}$ are independent. The likelihood function is then given by:

$$p\left(\vec{X}\middle|\vec{\Theta}\right) = \prod_{i=1}^{N} \frac{1}{(2\pi)^{1/2}\sigma} e^{\frac{-(x_i-\mu-\beta)^2}{2\sigma^2}} \tag{9.34}$$

We assume that we have no prior knowledge of the values of μ and σ and that β is normal, and as a result, the prior is given by

$$h\left(\vec{\Theta}\right) = \frac{1}{\sigma}\frac{1}{(2\pi)^{1/2}b} e^{\frac{-\beta^2}{2b^2}}. \tag{9.35}$$

Equations 9.34 and 9.35 can be combined to yield:

$$p\left(\vec{\Theta}\middle|\vec{X}\right) = \frac{c}{(2\pi)^{(N+1)/2}\sigma^{N+1}b} e^{-\frac{\sum_{i=1}^{N}(x_i-\mu-\beta)^2}{2\sigma^2} - \frac{\beta^2}{2b^2}} \tag{9.36}$$

Marginal pdfs (see the Appendix) for μ and σ are given, respectively, by:

$$p(\mu) = \int_{0}^{\infty}\int_{-\infty}^{\infty} p\left(\vec{\Theta}\middle|\vec{X}\right) d\beta d\sigma \tag{9.37}$$

$$p(\sigma) = \int_{-\infty}^{\infty}\int_{-\infty}^{\infty} p\left(\vec{\Theta}\middle|\vec{X}\right) d\beta d\mu \tag{9.38}$$

It is possible to analytically evaluate the integrals in Equations 9.37 and 9.38 for this particular example, but for other cases, this may not be possible and numerical integration may be the only way to make progress. To allow for more generality, we will evaluate Equations 9.37 and 9.38 numerically.

The following R code evaluates Equation 9.36 for discrete values of μ, σ, and β and then calculates a value of the constant c:

```
> N <- 10
> b <- 1
> x <- rnorm(n=N,mean=10,sd=1)
> mu <- seq(from=6,to=14,length.out=101)
> sigma <- seq(from=0.01,to=3,length.out=101)
> beta <- seq(from=-4,to=4,length.out=101)
> f <- array(0,c(length(mu),length(sigma),length(beta)))
> for (i in 1:length(mu)) {
+       for (j in 1:length(sigma)) {
+               for (k in 1:length(beta)) {
+                       h <- 1/sigma[j]
+                       f_x <- prod(dnorm(x=x-beta[k],mean=mu[i],sd=sigma[j]))
+                       f_beta <- dnorm(x=beta[k],mean=0,sd=b)
+                       f[i,j,k] <- f_x*f_beta*h
+               }
+       }
+ }
> # calculate a value of c via numerical integration
> dmu <- (max(mu)-min(mu))/(length(mu)-1)
> dsigma <- (max(sigma)-min(sigma))/(length(sigma)-1)
> dbeta <- (max(beta)-min(beta))/(length(beta)-1)
> c <- 1/sum(f*dmu*dsigma*dbeta,na.rm=TRUE)
> P <- c*f                      # this is Eq. (9.31)
```

In the code, we use a set of $N = 10$ data values and the values of μ, σ, and β are evaluated using nested for() loops. A simple integration scheme, where we summed the contributions from discrete volume elements, was used to evaluate the constant c. The value of $p\left(\vec{\Theta} \mid \vec{X}\right)$ in each volume element was assumed to be constant.

Here is R code to numerically evaluate the integrals in Equations 9.37 and 9.38 and plot the results:

```
> # plot the marginal pdf for mu
> p_mu_bayes <- numeric(length(mu))
> for (i in 1:length(mu)) {
+       p_mu_bayes[i] <- sum(P[i,,]*dsigma*dbeta)
+ }
> quartz(width=5.5,height=5.5)
> plot(mu,p_mu_bayes,type="l",xlab="mu",ylab="p_mu_bayes")
> # plot the marginal pdf for sigma
> p_sigma_bayes <- numeric(length(sigma))
> for (j in 1:length(sigma)) {
+       p_sigma_bayes[j] <- sum(P[,j,]*dmu*dbeta)
+ }
> quartz(width=5.5,height=5.5)
> plot(sigma,p_sigma_bayes,type="l",xlab="sigma",ylab="p_sigma_bayes")
```

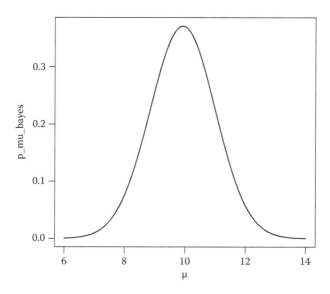

FIGURE 9.2
Marginal pdf for μ.

The marginal pdfs for μ and σ are shown in Figures 9.2 and 9.3, respectively. The plot in Figure 9.2 shows a symmetric pdf for μ, while Figure 9.3 shows that the pdf for σ is asymmetric.

The maximum a posteriori (MAP) estimates for the mean and standard deviation correspond to the peaks of each curve in Figures 9.2 and 9.3,

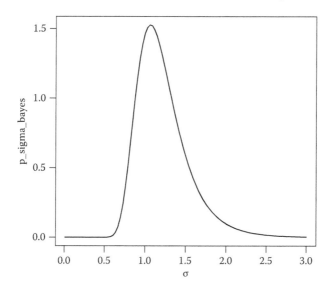

FIGURE 9.3
Marginal pdf for σ.

respectively. These values are obtained from the marginal pdfs using the following R code:

```
> # identify the MAPs for mu and sigma from the marginal pdfs
> sigma [which (p_sigma_bayes ==max (p_sigma_bayes))]
[1] 1.0565
> mu [which (p_mu_bayes==max (p_mu_bayes))]
[1] 9.92
```

The MAPs are $\mu = 9.92$ and $\sigma = 1.06$.

The marginal pdfs can also be used to estimate uncertainty intervals. We
accomplish this by integrating the marginal pdfs to form cumulative distribution functions (cdfs). The cdfs are plotted and the 95% uncertainty intervals for μ and σ are determined using the cdfs for these variables. This is
accomplished with the following R code:

```
> # create and plot a cdf for mu
> cdfmu <- numeric (length (mu))
> sum_mu <- 0
> for (i in 2:(length(mu))) {
+       sum_mu <- sum_mu+p_mu_bayes [i-1]*dmu
+       cdfmu [i] <- sum_mu
+ }
> quartz (width=5.5,height=5.5)
> plot (mu,cdfmu,type='l')
> # create and plot a cdf for sigma
> dsigma <- (max (sigma)-min (sigma))/(length (sigma)-1)
> cdfsigma <- numeric (length (sigma))
> sum_sigma <- 0
> for (i in 2:(length(sigma))) {
+       sum_sigma <- sum_sigma +p_sigma_bayes [i-1]*dsigma
+       cdfsigma [i] <- sum_sigma
+ }
> quartz (width=5.5,height=5.5)
> plot (sigma,cdfsigma,type='l')
> # calculate the mu mean and uncertainty
> conf_level <- 0.95
> cmin <- (1-conf_level)/2
> cmax <- (1+conf_level)/2
> I1 <- max (which (cdfmu<cmin))
> I2 <- min (which (cdfmu>cmax))
> mu_min <- (mu [I1]+mu [I1+1])/2
> mu_max <- (mu [I2]+mu [I2-1])/2
> mu_min
[1] 7.88
> mu_max
[1] 12.04
> # calculate the sigma mean and uncertainty
> J1 <- max (which (cdfsigma<cmin))
```

```
> J2 <- min(which(cdfsigma>cmax))
> sigma_min <- (sigma[J1]+ sigma[J1+1])/2
> sigma_max <- (sigma[J2]+ sigma[J2-1])/2
> sigma_min
[1] 0.77245
> sigma_max
[1] 2.02825
```

The cdfs for μ and σ are shown in Figures 9.4 and 9.5, respectively. The plot in Figure 9.4 is symmetric, but that in Figure 9.5 is not, which is reasonable given the shapes of the plots in Figures 9.2 and 9.3.

The 95% uncertainty intervals for μ and σ are calculated to be $9.48 \leq \mu \leq 11.08$ and $0.77 \leq \sigma \leq 2.03$. It is noted that we used only basic integration techniques here. More advanced integration methods could be used, but regardless of the integration method, the discretizations for μ, b, and σ should be varied to assess the accuracy of the numerical results.

Finally, we note that if we had more elemental systematic errors, then the computational resources required for the integrations can become substantial. For example, suppose we have L elemental systematic errors (β_i) corresponding to a data set with N measured values. Also assume that σ and μ are unknown. To evaluate a marginal pdf for μ, we would need to integrate over all possible β_i and σ values, which requires integrating over a space of $L + 1$ dimensions. If the β_i and σ values are each discretized into D subintervals (between finite limits of integration), then the total number of individual

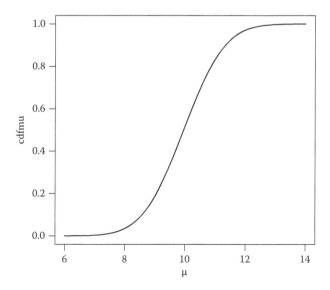

FIGURE 9.4
Cumulative density function plot for μ.

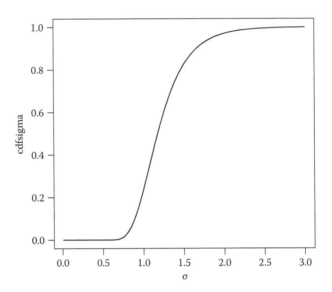

FIGURE 9.5
Cumulative density function plot for σ.

values that would need to be calculated to evaluate $p\left(\vec{\Theta}\middle|\vec{X}\right)$ at each discrete point in the space is D^{L+1}.

Suppose that $D = 100$ and $L = 10$ and then $D^{L+1} = 10^{22}$. This is a large number of values, and even if you had a computer that could store this many numbers, it would take a *very* long time just to calculate the 10^{22} values even once. For example, if you have a computer that calculates 10^{10} such values every second, it would take about 37000 years to calculate all of these values. As a result, alternative numerical integration methods are required for problems of even moderate complexity if a Bayesian approach is used. Markov chain Monte Carlo (MCMC) methods, which involve repeated sampling, have proven to be useful in this regard. The MCMC methodology is not covered here, but the interested reader can consult the references, e.g., [3,6], for more information.

Problems

P9.1 A sample space is composed of seven elementary events, i.e., $\Omega = \{X_1, X_2, X_3, X_4, X_5, X_6, X_7\}$. The events A, B, and C are $A = \{X_1, X_3, X_5, X_7\}$, $B = \{X_1, X_2, X_4, X_6, X_7\}$, and $C = \{X_1, X_4, X_6\}$. Calculate the following probabilities: $P(A)$, $P(C)$, $P(A \cup B)$, $P(C \cap B)$, $P(A \mid B)$, and $P(A \mid C)$.

P9.2 A family has two children. One of them is a girl. What is the probability that the other child is a boy?

P9.3 The ACME corporation manufactures widgets, and it is known that 0.001% of widgets produced are defective. A special machine (a widget tester) tests each widget to determine whether it is defective. The widget tester is 99% accurate; i.e., it correctly flags widgets as being defective 99% of the time. If a widget is flagged as defective, what is the probability that it is actually defective?

P9.4 Repeat P9.3 but for the case where a second widget tester is used to test the widgets that the first tester flagged as being defective. If a widget is flagged as defective by the second widget tester, what is the probability that it is actually defective?

P9.5 Redo the analysis that produced Figures 9.3 and 9.4 but with the priors $h(\vec{\Theta}) = 1$ and $h(\vec{\Theta}) = e^{-\sigma}$. Also vary N and comment on whether the choice of prior has a significant influence for large N.

P9.6 Generate simulated data using the following R code:

```
x <- rexp(n=N,rate=0.5)
```

Assume that the data you generate are described by the pdf $p(x) = \lambda \exp(-\lambda x)$, $x_{min} = 0$, $x_{max} = \infty$. Plot the MAP for λ vs. N for values of N generated with `N <- seq(from=2,to=1e5,by=1)`. Use a Bayesian approach, and for simplicity, use the prior $h(\vec{\Theta}) = 1/\lambda$.

P9.7 Generate simulated data using the following R code:

```
x <- rnorm(n=N,mean=5,sd=1/sqrt(2))
```

Assuming that there is no uncertainty in σ, i.e., $\sigma = 1/\sqrt{2}$, determine the 95% uncertainty interval for μ using a Bayesian approach for $N = \{10, 20, 50, 100\}$. For simplicity, use the prior $h(\vec{\Theta}) = 1$.

P9.8 Generate simulated data using the following R code:

```
x <- rnorm(n=N,mean=0,sd=2)
```

Assuming that there is no uncertainty in μ, i.e., $\mu = 0$, determine the 95% uncertainty interval for σ using a Bayesian approach for $N = \{10, 20, 50, 100\}$. For simplicity, use the prior $h(\vec{\Theta}) = 1/\sigma$.

P9.9 Derive analytical expressions for the MAPs of μ and σ for the problem corresponding to Figure 9.2.

P9.10 Repeat the analysis that produced Figures 9.2 through 9.5, but instead use a symmetric triangular pdf for β with a mean of zero and a standard deviation of unity.

References

1. F. James, *Statistical Methods in Experimental Physics*, 2nd edn., World Scientific, Hackensack, NJ, 2006.
2. H. J. C. Berendsen, *A Student's Guide to Data and Error Analysis*, Cambridge University Press, New York, 2011.
3. N. T. Hobbs and M. B. Hooten, *Bayesian Models—A Statistical Primer for Ecologists*, Princeton University Press, Princeton, NJ, 2015.
4. J. V. Stone, *Bayes' Rule with R: A Tutorial Introduction to Bayesian Analysis (Tutorial Introductions)*, Vol. 5, Sebtel Press, Sheffield, U.K., 2016.
5. A. Possolo and B. Toman, Tutorial for metrologists on the probabilistic and statistical apparatus underlying the GUM and related documents, National Institute of Standards and Technology, Gaithersburg, MD, 2011, www.itl.nist.gov/div898/possolo/TutorialWEBServer/TutorialMetrologists2011Nov09.xht. (accessed December 31, 2016).
6. M. J. Crawley, *The R Book*, 2nd edn., Wiley, West Sussex, U.K., 2013.
7. P. H. Garthwaite, J. B. Kadane, and A. O'Hagan, Statistical methods for eliciting probability distributions, *Journal of the American Statistical Association* 100, 680–701, 2005.
8. H. Jeffreys, *Theory of Probability*, 3rd edn., Oxford University Press, London, U.K., 1961.
9. J. Bernardo and A. Smith, *Bayesian Theory*, 2nd edn., Wiley, Chichester, England, 2007.

Appendix: Probability Density Functions

The concept of a probability density function (pdf) is important in data analysis [1–3]. Probability density functions (pdfs) show up in a natural fashion, e.g., when uncertainty intervals are evaluated. Here, we consider some basic aspects of pdfs for both univariate and multivariate cases, i.e., when we have only one or more than one random variable. We start with a discussion of univariate pdfs.

A.1 Univariate pdfs

A univariate pdf is an equation, denoted here as $p(x)$, that generally has its greatest relevance when it is integrated. The symbol x represents a single random variable. A pdf is never negative and the integral of a pdf between two values x_1 and x_2 gives the probability $P_{x_1 \to x_2}$ that a measurement will fall within the range x_1–x_2:

$$P_{x_1 \to x_2} = \int_{x_1}^{x_2} p(x)dx. \tag{A.1}$$

Note that a pdf is also termed a distribution.

The cumulative distribution function (cdf) is found by integrating from the minimum possible data value, x_{min}, to some value of x [4]:

$$\mathrm{cdf}(x) = \int_{x_{min}}^{x} p(x)dx = P_{x_{min} \to x}. \tag{A.2}$$

The derivative of a cdf with respect to x is the original pdf, i.e.,

$$\frac{d}{dx}\mathrm{cdf}(x) = p(x). \tag{A.3}$$

The cdf is the probability that a measured value is less than or equal to x. If we integrate from x_{min} to the maximum possible data value, x_{max}, the integral is unity,

$$P_{x_{min} \to x_{max}} = \int_{x_{min}}^{x_{max}} p(x)dx = 1, \qquad (A.4)$$

because a measurement must produce *some* possible value.

The mean μ, variance σ^2, and standard deviation σ are defined as

$$\mu = \int_{x_{min}}^{x_{max}} xp(x)dx, \qquad (A.5)$$

$$\sigma^2 = \int_{x_{min}}^{x_{max}} (x-\mu)^2 p(x)dx, \qquad (A.6)$$

and

$$\sigma = \sqrt{\sigma^2}. \qquad (A.7)$$

These expressions are general in the sense that they apply to any univariate pdf. There are only a few such pdfs, however, that are typically important for the analysis of experimental data: normal (Gaussian), uniform, triangular, Student's t, and chi-square. We discuss these briefly in the following.

A.1.1 Normal Distribution

The normal distribution is arguably the most important distribution in data analysis. This is because, e.g., it shows up as a consequence of the central limit theorem [1,3], which states that the distribution of a random variable such as the mean of a sample is normal as the number of samples goes to infinity [5]. The normal distribution is defined as

$$p(x) = \frac{1}{\sigma\sqrt{2\pi}} e^{\frac{-(x-\mu)^2}{2\sigma^2}}, \quad x_{min} = -\infty, x_{max} = +\infty. \qquad (A.8)$$

The values for x_{min} and x_{max} may not be physically attainable, but it is mathematically convenient to use them and it is often the case that they are a reasonable approximation. Some important functions in R are as follows:

```
> dnorm(x=0.5,mean=1,sd=2)      # value of p(x)
[1] 0.1933341
```

```
> pnorm(q=0.5,mean=1,sd=2)      # cdf value
[1] 0.4012937
> qnorm(p=0.5,mean=1,sd=2)      # inverse cdf
[1] 1
```

Note that the pattern in R is to use the first letter in each of these functions to denote the quantity of interest and the rest of the letters to denote the distribution. In particular, "d" is density (the value of the pdf for a given x value), "p" is cumulative probability (the cdf value for a given value of x), and "q" is the quantile for a given cdf value; e.g., qnorm(p=0.5,mean=1,sd=2) gives the value of x that corresponds to a cdf of 0.5 with a mean of 1 and a standard deviation of 2 for a normal distribution. The quantile is the inverse cdf.

The normal distribution plots out as the famous bell-shaped curve, as shown in Figure A.1 for $\mu = 1$ and $\sigma = 2$. The R code to generate this plot is as follows:

```
> curve(dnorm(x, mean=1,sd=2),xlim=c(-6,8))
```

The cdf for a normal distribution is a sigmoidal (S-shaped) curve, as shown in Figure A.2 for $\mu = 1$ and $\sigma = 2$. The R code to generate this plot is

```
> curve(pnorm(x, mean=1,sd=2),xlim=c(-6,8))
```

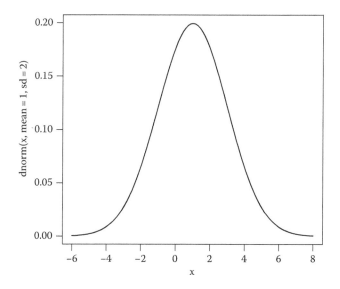

FIGURE A.1
Normal pdf with a mean of 1 and a standard deviation of 2.

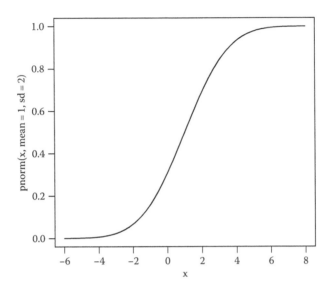

FIGURE A.2
Normal cdf with a mean of 1 and a standard deviation of 2.

It is sometimes convenient to express the normal distribution in terms of the "z" variable:

$$z = \frac{x - \mu}{\sigma}. \tag{A.9}$$

The pdf for z is then given by

$$p(z) = \frac{1}{\sqrt{2\pi}} e^{\frac{-z^2}{2}}, \quad z_{min} = -\infty, z_{max} = +\infty. \tag{A.10}$$

The relevant z-related functions in R are as follows (where, by just omitting the mean and standard deviation arguments, R performs the calculations for z):

```
> dnorm(0.5)        # value of p(z)
[1] 0.3520653
> pnorm(q=0.5)      # cdf value
[1] 0.6914625
> qnorm(p=0.5)      # inverse cdf
[1] 0
```

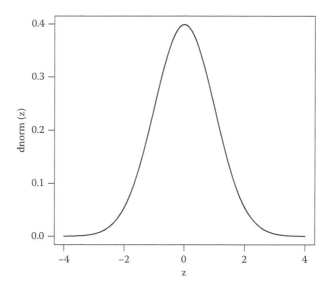

FIGURE A.3
Normal pdf with a mean of 0 and a standard deviation of 1.

Figures A.3 and A.4 show pdf and cdf plots in terms of z. Note that the pdf for z is the normal distribution with $\mu = 0$ and $\sigma = 1$. These plots are obtained with the following code:

```
> curve(dnorm(x),xlim=c(-4,4),ylab="dnorm(z)",xlab="z")
> curve(pnorm(x),xlim=c(-4,4),ylab="pnorm(z)",xlab="z")
```

A.1.2 Uniform Distribution

The uniform distribution is used when a random variable (x) is constrained to be within the range $x_{min} \leq x \leq x_{max}$, but the probability of a value falling within a subrange Δx is the same regardless of where Δx is located. The uniform distribution is defined as

$$p(x) = \begin{cases} \dfrac{1}{x_{max} - x_{min}}, & x_{min} \leq x \leq x_{max}, \\ 0, & \text{otherwise.} \end{cases} \qquad (\text{A.11})$$

Some important functions in R are as follows:

```
> dunif(x=1,min=0,max=2)    # value of p(x)
[1] 0.5
```

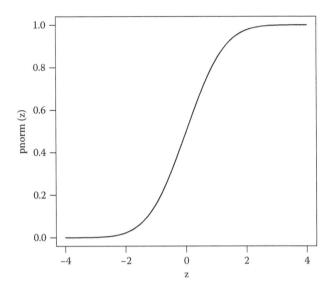

FIGURE A.4
Normal cdf with $\mu = 0$ and $\sigma = 1$.

```
> punif(q=1,min=0,max=2)    # cdf value
[1] 0.5
> qunif(p=0.5,min=0,max=2) # inverse cdf
[1] 1
```

This distribution has the shape of a rectangle and the cdf is a ramp (Figures A.5 and A.6). Here is the R code to generate the plots:

```
> curve(dunif(x,min=0,max=2),xlim=c(-1,3),n=501)
> curve(punif(x,min=0,max=2),xlim=c(-1,3),n=501)
```

The mean, variance, and standard deviation for a uniform distribution are

$$\mu = \frac{x_{min} + x_{max}}{2}, \tag{A.12}$$

$$\sigma^2 = \frac{\left(x_{max} - x_{min}\right)^2}{12}, \tag{A.13}$$

and

$$\sigma = \frac{\left(x_{max} - x_{min}\right)}{\sqrt{12}}. \tag{A.14}$$

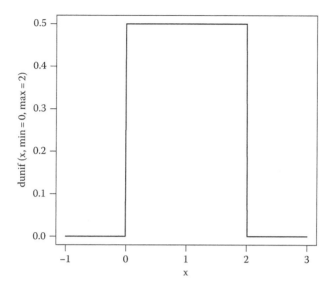

FIGURE A.5
Uniform pdf with $x_{min} = 0$ and $x_{max} = 2$.

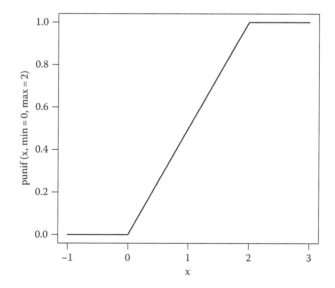

FIGURE A.6
A uniform distribution's cdf with $x_{min} = 0$ and $x_{max} = 2$.

A.1.3 Triangular Distribution

The triangular distribution applies to a situation in which $x_{min} \leq x \leq x_{max}$, but where the pdf has a peak such that an experimental measurement is more likely to produce a result near the peak than away from it. The triangular distribution is defined as

$$p(x) = \begin{cases} \dfrac{2}{x_{max} - x_{min}} \dfrac{x - x_{min}}{x_{peak} - x_{min}}, & x_{min} \leq x < x_{peak}, \\[3mm] \dfrac{2}{x_{max} - x_{min}}, & x = x_{peak}, \\[3mm] \dfrac{2}{x_{max} - x_{min}} \dfrac{x - x_{max}}{x_{peak} - x_{max}}, & x_{peak} < x \leq x_{max}, \\[3mm] 0, & \text{otherwise}, \end{cases} \tag{A.15}$$

where x_{peak} is the x value corresponding to the top of the triangle.
    Some important functions in R are as follows:

```
> library(triangle)                    # load the library "triangle"
> dtriangle(x=1,a=0,b=2,c=1)           # value of p(x)
[1] 1
> ptriangle(q=1,a=0,b=2,c=1)           # cdf value
[1] 0.5
> qtriangle(p=0.5,a=0,b=2,c=1)         # inverse cdf
[1] 1
```

Note that the argument lists use $x_{min} = a$, $x_{max} = b$, and $x_{peak} = c$. This distribution has the shape of a triangle (Figure A.7) and the cdf is sigmoidal (Figure A.8). Here is the code for the plots:

```
> curve(dtriangle(x,a=0,b=2,c=1),xlim=c(-1,3),n=501)
> curve(ptriangle(x,a=0,b=2,c=1),xlim=c(-1,3),n=501)
```

The mean, variance, and standard deviation for a triangle distribution are

$$\mu = \frac{x_{min} + x_{peak} + x_{max}}{3}, \tag{A.16}$$

$$\sigma^2 = \frac{x_{min}^2 + x_{peak}^2 + x_{max}^2 - x_{min}x_{max} - x_{min}x_{peak} - x_{max}x_{peak}}{18}, \tag{A.17}$$

and

$$\sigma = \sqrt{\frac{x_{min}^2 + x_{peak}^2 + x_{max}^2 - x_{min}x_{max} - x_{min}x_{peak} - x_{max}x_{peak}}{18}}. \tag{A.18}$$

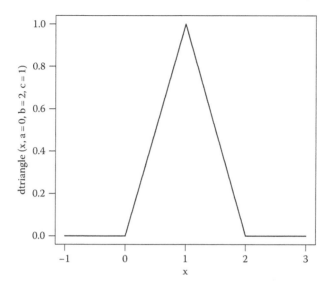

FIGURE A.7
Triangle pdf for $x_{min} = 0$, $x_{max} = 2$, and $x_{peak} = 1$.

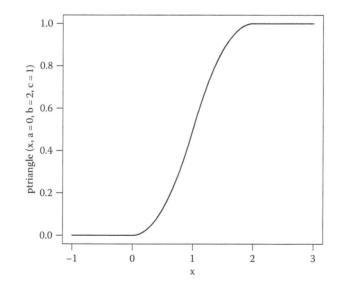

FIGURE A.8
A triangle distribution's cdf for $x_{min} = 0$, $x_{max} = 2$, and $x_{peak} = 1$.

If the triangle is *symmetric*, i.e., $x_{peak} = (x_{min} + x_{max})/2$, these expressions simplify to

$$\mu = \frac{x_{min} + x_{max}}{2},$$ (A.19)

$$\sigma^2 = \frac{(x_{max} - x_{min})^2}{24},$$ (A.20)

and

$$\sigma = \frac{(x_{max} - x_{min})}{\sqrt{24}}.$$ (A.21)

A.1.4 Student's *t* Distribution

The *t* distribution is used for the estimation of confidence intervals of the mean when the standard deviation is unknown (which is generally the case). Student's *t* distribution is defined as

$$p(t) = \frac{1}{\sqrt{v\pi}} \frac{\Gamma\left(\dfrac{v+1}{2}\right)}{\Gamma\left(\dfrac{v}{2}\right)} \left(1 + \frac{t^2}{v}\right)^{\frac{-(v+1)}{2}}, \quad t_{min} = -\infty, t_{max} = +\infty,$$ (A.22)

where Γ denotes the gamma function [6] and the *t* variable is defined as

$$t = \frac{\bar{x} - \mu}{S/\sqrt{N}}.$$ (A.23)

Some important functions in R are as follows:

```
> dt (x=1,df=9)        # value of p(x)
[1] 0.2291307
> pt (q=1,df=9)        # cdf value
[1] 0.8282818
> qt (p=0.5,df=9)      # inverse cdf
[1] 0
```

The *t* distribution is symmetric about $t = 0$ and is similar to the normal distribution but with a wider distribution and longer "tails" on either side of the mean. The *t* distribution approaches the normal distribution as v increases and the two distributions are nearly identical for $v > 30$. The cdf for this

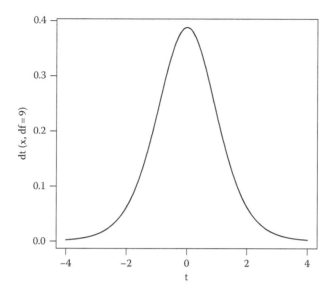

FIGURE A.9
Student's t pdf for $\nu = 9$.

distribution is sigmoidal. Representative plots for the t pdf and cdf, shown in Figures A.9 and A.10, respectively, are generated with the following code:

```
> curve(dt(x,df=9),xlim=c(-4,4),n=501,xlab='t')
> curve(pt(x,df=9),xlim=c(-4,4),n=501,xlab='t')
```

The variance and standard deviation for the t distribution are

$$\sigma^2 = \frac{\nu}{\nu - 2} \qquad (A.24)$$

and

$$\sigma = \sqrt{\frac{\nu}{\nu - 2}}. \qquad (A.25)$$

The variance and standard deviation are defined only for $\nu > 2$. The variable ν is termed the "number of degrees of freedom."

A.1.5 Chi-Square Distribution

The chi-square distribution is important when dealing with a sum of squares of normal random variables. An example would be the chi-square goodness-of-fit test. The chi-square distribution is defined as

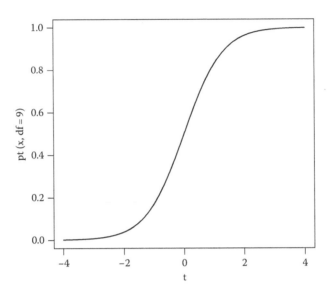

FIGURE A.10
A Student's *t* distribution cdf for $\nu = 9$.

$$p(x) = \frac{\dfrac{1}{2}\left(\dfrac{x}{2}\right)^{(\nu/2)-1}}{\Gamma\left(\dfrac{\nu}{2}\right)} e^{-x/2}, \quad x_{min} = 0, x_{max} = +\infty, \tag{A.26}$$

where ν is the number of degrees of freedom.
    Some important functions in R are as follows:

```
> dchisq(x=1,df=10)          # value of p(x)
[1] 0.0007897535
> pchisq(q=1,df=10)          # cdf value
[1] 0.0001721156
> qchisq(p=0.5,df=10)        # inverse cdf
[1] 9.341818
```

The chi-square distribution is asymmetric about the mean, but it approaches the normal distribution as ν increases. The cdf for this distribution is sigmoidal. Representative plots of the chi-square pdf and cdf, shown in Figures A.11 and A.12, respectively, are generated with the following code:

```
> curve(dchisq(x,df=10),xlim=c(0,30),n=501)
> curve(pchisq(x,df=10),xlim=c(0,30),n=501)
```

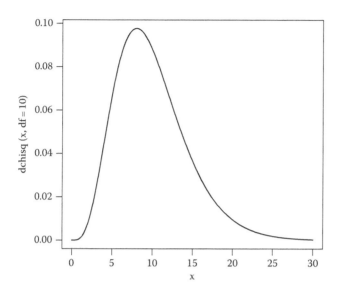

FIGURE A.11
Chi-square pdf for $\nu = 10$.

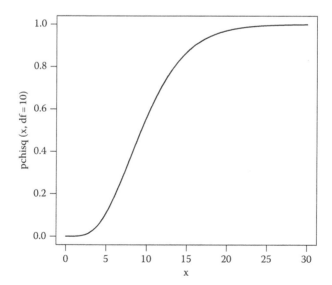

FIGURE A.12
A chi-square distribution's cdf for $\nu = 10$.

The mean, variance, and standard deviation for the chi-square distribution are

$$\mu = \nu, \tag{A.27}$$

$$\sigma^2 = 2\nu, \tag{A.28}$$

and

$$\sigma = \sqrt{2\nu}. \tag{A.29}$$

A.2 Multivariate pdfs

A multivariate pdf is a function of several random variables $x_1, x_2, ..., x_L$ such that the probability dP that a set of measured values $x_1, x_2, ..., x_L$ will be in the ranges $x_1 \rightarrow x_1 + dx_1, x_2 \rightarrow x_2 + dx_2, ..., x_L \rightarrow x_L + dx_L$ is

$$dP = p(x_1, x_2, ..., x_L) dx_1 dx_2 \cdots dx_L. \tag{A.30}$$

For brevity, we will use the notation

$$p(x_1, x_2, ..., x_L) = p(\vec{x}) \tag{A.31}$$

and

$$dx_1 dx_2 \cdots dx_L = d\vec{x}. \tag{A.32}$$

Equation A.30 then can be written as

$$dP = p(\vec{x}) d\vec{x}. \tag{A.33}$$

When integrated over all possible values of $\vec{x}$, Equation A.33 will integrate to unity:

$$\int_{\text{All } \vec{x}} p(\vec{x}) d\vec{x} = 1. \tag{A.34}$$

In Equation A.34, the single integral symbol implies that integrals are to be performed over every random variable.

The variance of a function $r(\bar{x})$ is given by

$$V(r) = \langle r^2 \rangle - \langle r \rangle^2,$$ (A.35)

where the mean values $\langle r \rangle$ and $\langle r^2 \rangle$ are given by

$$\langle r \rangle = \int_{\text{All } \bar{x}} r(\bar{x}) p(\bar{x}) d\bar{x}$$ (A.36)

and

$$\langle r^2 \rangle = \int_{\text{All } \bar{x}} r^2(\bar{x}) p(\bar{x}) d\bar{x}.$$ (A.37)

A.3 Marginal Distributions

Marginal distributions are the pdfs we are left with after integrating a multivariate pdf over some of the random variables [7]. For example, consider a bivariate pdf $p(x_1, x_2)$. The marginal distribution for x_1, which we denote as M_1, is given by integrating over all possible values of x_2 as follows:

$$M_1 = \int_{x_{2,\text{min}}}^{x_{2,\text{max}}} p(x_1, x_2) dx_2.$$ (A.38)

In the special case where $p(x_1, x_2)$ is a product of two pdfs,

$$p(x_1, x_2) = p_1(x_1) p_2(x_2),$$ (A.39)

then Equation A.38 can be written as

$$M_1 = \phi_1 p_1(x_1),$$ (A.40)

where the variable ϕ_1 is defined as

$$\phi_1 = \int_{x_{2,\text{min}}}^{x_{2,\text{max}}} p_2(x_2) dx_2.$$ (A.41)

However, $p_2(x_2)$, when integrated over all possible values of x_2, is unity. We then have $\phi_1 = 1$ such that

$$M_1 = p_1(x_1) \tag{A.42}$$

holds. The marginal distribution for x_2, which we denote as M_2, can be similarly evaluated to yield

$$M_2 = p_2(x_2). \tag{A.43}$$

References

1. JCGM 100:2008, Evaluation of measurement data—Guide to the expression of uncertainty in measurement, GUM 1995 with minor corrections, International Bureau of Weight and Measures (BIPM), Sérres, France 2008.
2. JCGM 101:2008, Evaluation of measurement data: Supplement 1 to the "Guide to the expression of uncertainty in measurement"—Propagation of distributions using a Monte Carlo method, International Bureau of Weight and Measures (BIPM), Sérres, France 2008.
3. H. W. Coleman and W. G. Steele, *Experimentation, Validation, and Uncertainty Analysis for Engineers*, 3rd edn., Wiley, Hoboken, NJ, 2009.
4. N. Radziwill, *Statistics (the Easier Way) with R: An Informal Text on Applied Statistics*, Lapis Lucera, San Francisco, CA, 2015.
5. T. A. Garrity, *All the Mathematics You Missed: But Need to Know for Graduate School*, Cambridge University Press, New York, 2001.
6. C. M. Bender and S. A. Orszag, *Advanced Mathematical Methods for Scientists and Engineers*, McGraw-Hill, New York, 1978.
7. N. T. Hobbs and M. B. Hooten, *Bayesian Models—A Statistical Primer for Ecologists*, Princeton University Press, Princeton, NJ, 2015.

Index